CRYSTAL CHEMISTRY OF LARGE-CATION SILICATES

KRISTALLOKHIMIYA SILIKATOV S KRUPNYMI KATIONAMI

КРИСТАЛЛОХИМИЯ СИЛИКАТОВ С КРУПНЫМИ КАТИОНАМИ

CRYSTAL CHEMISTRY of LARGE-CATION SILICATES

by

Acad. N. V. BELOV

Authorized translation from the Russian

Springer Science+Business Media, LLC

ACKNOWLEDGMENT

The Publisher wishes to thank the copyright owner, the American Institute of Physics, Inc., for permission to reprint the following papers which originally appeared in *Soviet Physics — Doklady* or *Soviet Physics — Crystallography*.

Nikolai Vasil'evich Belov (On His Seventieth Birthday), *Sov. Phys. Cryst.* Vol. 6, pp. 661-664; The Crystal Structure of Lawsonite, *Sov. Phys. Dokl.* Vol. 4, pp. 20-23; The Determination of the Structure of Seidozerite, *Sov. Phys. Cryst.* Vol. 4, pp. 146-157; The Structures of Herderite, Datolite, and Gadolinite Determined by Direct Methods, *Sov. Phys. Cryst.* Vol. 4, pp. 300-314; Derivation of the Structure of Lovozerite from Sections of the Three-Dimensional Patterson Function, *Sov. Phys. Cryst.* Vol. 5, pp. 186-198; Baotite — A Mineral with $[Si_4O_{12}]$ Metasilicate Rings, *Sov. Phys. Cryst.* Vol. 5, pp. 523-525; Crystal Structure of Narsarsukite *Sov. Phys. Cryst.* Vol. 5, pp. 540-548; Crystal Structure of Spurrite, *Sov. Phys. Cryst.* Vol. 5, pp. 659-667; The Crystal Structure of Hurlbutite, *Sov. Phys. Dokl.* Vol. 5, pp. 1141-1144; Crystal Structure of Paracelsian, *Sov. Phys. Cryst.* Vol. 5, pp. 826-829; The Crystal Structure of Rubidium Di(Meta)Fluoroberyllate $(RbBe_2F_5)$ and Its Relationship to Silicate Sheet Structures with $[Si_2O_5]$ Units, *Sov. Phys. Cryst.* Vol. 6, pp. 685-693.

The following papers appeared previously in *Doklady of the Academy of Sciences of the USSR, Earth Sciences Sections*, published by the American Geological Institute:

The Crystalline Structure of Tricalcium Silicate Hydrate (Vol. 126, pp. 574-576); Crystal Structure of Lovenite (Vol. 130, pp. 167-170).

The following papers appeared previously in *Proceedings of the Academy of Sciences of the USSR, Geological Sciences Sections* or *Journal of Structural Chemistry*, both published by Consultants Bureau Enterprises, Inc.:

Crystal Structure of Mica-like Hydrous Calcium Silicates: Okenite, Nekoite, Truscottite, and Gyrolite. New Silicate Radical $[Si_6O_{15}]_\infty$, *Proc. Acad. Sci. USSR*, Vol. 121, pp. 713-716; The Crystal Structure of Foshagite — $Ca_8[Si_6O_{17}](OH)_6$, *Proc. Acad. Sci. USSR*, Vol. 121, pp. 725-727; The Crystalline Structure of Tobermorites, *Proc. Acad. Sci. USSR*, Vol. 123, pp. 1035-1037; The Structure of Epididymite $NaBeSi_3O_7(OH)$. A New Type of $[Si_6O_{15}]$ Chain, *J. Struct. Chem.* Vol. 1, pp. 44-54.

The original Russian text was published in 1961 by The Academy of Sciences Press, in Moscow, for the Institute of Geochemistry and Analytical Chemistry of the Academy of Sciences of the USSR, as the Second V. I. Vernadskii Lecture.

Кристаллохимия силикатов с крупными катионами

Николай Васильевич Белов

Library of Congress Catalog Card Number 63-17642

ISBN 978-1-4899-4607-2   ISBN 978-1-4899-4605-8 (eBook)
DOI 10.1007/978-1-4899-4605-8

CONTENTS

CRYSTAL CHEMISTRY OF LARGE-CATION SILICATES

V. I. Vernadskii and Silicate Crystal Chemistry 1
The Second "Chapter" of Silicate Crystal Chemistry. 3
Isomorphism Relations Between Zirconium and Titanium 13
Oxygen-Silicon Chains and Ribbons in the Second "Chapter" of Crystal Chemistry
     of Silicates 14
The Crystal Structure of Rhodonite—$MnSiO_3$ 20
Oxygen—Silicon Networks Based on Wollastonite Chains and Xonotlite Ribbons 23
Molecular Sieves 27
Literature Cited. 45

APPENDIX

Nikolai Vasil'evich Belov (On His Seventieth Birthday) 49
Crystal Structure of Mica-Like Hydrous Calcium Silicates: Okenite, Nekoite, Truscottite
        and Gyrolite. New Silicate Radical $[Si_6O_{15}]_\infty$
     Kh. S. Mamedov and Academician N. V. Belov. 53
The Crystal Structure of Foshagite—$Ca_8[Si_6O_{17}](OH)_6$
     Kh. S. Mamedov and Academician N. V. Belov. 57
The Crystalline Structure of Tobermorites
     Kh. S. Mamedov and Academician N. V. Belov. 60
The Crystal Structure of Lawsonite
     I. M. Rumanova and T. I. Skipetrova 63
The Determination of the Structure of Seidozerite
     V. I. Simonov and N. V. Belov. 67
The Crystalline Structure of Tricalcium Silicate Hydrate
        $TSH = 6CaO \cdot 2SiO_2 \cdot 3H_2O = Ca_6[Si_2O_7](OH)_6 = Ca_4[Si_2O_7](OH)_2 \cdot 2Ca(OH)_2$
     Kh. S. Mamedov, R. F. Klevtsova, and Academician N. V. Belov 79
The Structures of Herderite, Datolite and Gadolinite Determined by Direct Methods
     P. V. Pavlov and N. V. Belov 82
Crystal Structure of Lovenite
     V. I. Simonov and Academician N. V. Belov. 97
Derivation of the Structure of Lovozerite from Sections of the Three-Dimensional
        Patterson Function
     V. V. Ilyukhin and N. V. Belov. 101
The Structure of Epididymite $NaBeSi_3O_7(OH)$. A New Type of $[Si_6O_{15}]$ Chain
     E. A. Pobedimskaya and N. V. Belov 114
Baotite—A Mineral with $[Si_4O_{12}]$ Metasilicate Rings
     V. I. Simonov 125
Crystal Structure of Narsarsukite
     Yu. A. Pyatenko and Z. V. Pudovkina 128
Crystal Structure of Spurrite
     R. F. Klevtsova and N. V. Belov 137

The Crystal Structure of Hurlbutite

      V. V. Bakakin and Academician N. V. Belov. 146

Crystal Structure of Paracelsian

      V. V. Bakakin and N. V. Belov. 150

The Crystal Structure of Rubidium Di(meta)fluoroberyllate (RbBe$_2$F$_5$) and Its Relationship

      to Silicate Sheet Structures with [Si$_2$O$_5$] Units

      V. V. Ilyukhin and N. V. Belov. 154

V. I. Vernadskii and Silicate Crystal Chemistry

While at St. Petersburg University, V. I. Vernadskii studied under three of the greatest Russian scientists of the day: Mendeleev, creator of the Periodic Table of Elements, Butlerov, one of the founders of modern organic chemistry, and Dokuchaev, the great soil mineralogist. After completing his university studies, young Vernadskii was undecided as to which of the three branches of science represented by each of these scientists he would devote himself. It is well known that he was attracted to organic chemistry, so when he had finally settled on mineralogy, Vernadskii rapidly and energetically set about transforming this into the chemistry of the earth's crust, in fact, into geochemistry. Using the periodic law, the young scientist established the mineralogical isomorphous series, which is still named for him.

The fundamental principle of organic chemistry is the existence of radicals, i.e., isolated groups of atoms which, while retaining their individuality inside an organic molecule, can be transferred unchanged to another molecule, and which act as the common denominator in series of substances, giving them similar overall properties. In the chemistry of the earth's crust, which is primarily silicate chemistry, it was much more difficult to unearth similar characteristic radicals because ordinary chemical methods, based on solution of the individual radicals without changing them, could not be used. So the mineralogist or geochemist could only study the chemical reactions occurring in the crystallization of rocks from magma, or in the accompanying metamorphic processes, from the reaction products produced by nature — minerals and their surroundings (paragenesis).

Nonetheless, Vernadskii very quickly (1891) managed to discover the basic radical which seemed to be present in a majority of the alumosilicates (the earth's crust consists of 50% oxygen, 25% silicon, 9% aluminum, and other elements, none of which makes up more than 3-4%). This radical was the famous "kaolin nucleus," and Vernadskii made the most of it to arrange almost all alumosilicates in a strict classification. Russian mineralogists have used this classification almost unaltered right up to the present, the only minor modification being the replacement of molecular representations of silicates by coordination models.

In spite of the difficulty of finding the basic radicals present in other silicates, Vernadskii foresaw that the problem would be solved in later years using the techniques of microcrystallography*and crystal chemistry, the theoretical foundations of which had just been laid by E. S. Fedorov. Experimental work in this direction became possible in 1912, when Laue discovered interference of X rays by crystals.

An untimely death overtook Fedorov (1919), and then also G. V. Vul'f (1925). This was a great setback in the development of silicate crystal chemistry in the Soviet Union. Scientific historians take delight in mentioning the very early views of Vul'f on the structure of mica, expressed in the early twenties, but it must be admitted that this material is of consequence only for memoirs. This very elegant idea, based on the epitaxial growth of KI on micas, resulted in correct values for the lattice pa-

---

*In the very early years of his teaching career, Vernadskii had devised an original crystallography course, in which some interesting ideas on the essence of polymorphism had been put forward.

rameters of mica. His proposals could not be confirmed experimentally because at that time structural analysis was only in its infancy.

Soviet silicate crystal chemistry began in 1934-35, when Fersman suggested that I prepare an enlarged Russian translation of Hassel's "Crystal Chemistry," which dealt comprehensively with silicates, and directly after this, the third volume of "Fundamental Ideas in Geochemistry," the principal contents of which included such important papers as Bragg's "Structure of Silicates" and Schiebold's work of the same title. At the same time Vernadskii proposed that I participate in the "crystal chemical" part of the work for his new book "Terrestrial Silicates and Alumosilicates" (the other author was S. M. Kurbatov), which was published in 1937.

Also in 1935, J. Bernal gave a course on crystal chemistry in Leningrad and Moscow, and much of this course was devoted to silicates. Even more important was the personal participation of Bernal in the first Soviet structural analysis of a silicate, katapleite (1935-36)[1], carried out in the X-ray department of our Academical Crystallography Laboratory by its founder and first chief, B. K. Brunovskii (a pupil of Vernadskii).

In the years which followed there was a long period of inactivity in the study of silicates, both in the Soviet Union and in the West, brought about by a number of causes. Above all, there existed the widespread view (or prejudice) that the chemistry of silicates, in particular natural silicates, was governed by a small number of laws, limited by only a very small number of different conceivable oxygen-silicon radicals; this was believed in spite of the fact that the minerals in question make up 95% of the earth's crust. It was considered that these radicals had been established once and for all by W. L. Bragg in Manchester, during 1926-31, and were standardized in his famous article "Structure of Silicates," which should possibly be supplemented by Pauling's structures of sheet silicates (micas, chlorites, and so on). It appeared certain that crystal chemistry of silicates was just a matter of finding one of the Bragg-Pauling radicals in each new silicate structure. The zirconium silicate katapleite, $Na_2Zr[Si_3O_9] \cdot 2H_2O$, turned out to be a lucky example. This was the second example of a silicate with a three-membered silicon oxygen ring, $[Si_3O_9]$, identical to that figuring in Bragg's titanium silicate benitoite, $BaTi[Si_3O_9]$.

Nowadays it is surprising that the structural classification of silicates, established by Bragg and Pauling and described in these two books, should have retained its importance as the sole reference source for more than 25 years, and be incorporated without any modifications or additions in all treatises on crystal chemistry and the appropriate sections of mineralogy (for example, the second edition of Winkler's "Crystal Chemistry," published in 1957, or Strunz's "Mineralogical Tables," also published in 1957). It is admittedly true that the small number of silicates for which structures have been determined since then have fitted into the framework of the standard quite well. In the Moscow Crystallography Institute, the structure of dioptase, $Cu_6[Si_6O_{18}] \cdot 6H_2O$, was worked out, and this turned out to be another example of what Bragg, Pauling, and Strunz called a sorosilicate, with six-membered oxygen-silicon rings [2]. Solved in the same institute, the structure of ramsayite, $Na_2Ti_2Si_2O_9$, presented a second example of a metasilicate (inosilicate), containing $[Si_2O_6]_\infty$ chains of Bragg's pyroxene type [3], with the important point that the empirical formula of this mineral was not of the meta type, but of the ortho type (O:Si = 4.5:1). At the same time, the fact was not overlooked that this structure represented an outstanding example of Vernadskii's "organic" principle; it showed the presence of a metasilicate radical $[SiO_3]_\infty$ within a more complex system of oxides; instead of the normal formula of $Na_2O \cdot 2TiO_2 \cdot 2SiO_2 = Na_2Ti_2Si_2O_9$, we now write the rational formula of ramsayite, extracting this radical from it, as $Na_2Ti_2[Si_2O_6]O_3$.

Rather more surprising was the discovery in milarite (the same institute [4]) of a new type of oxygen-silicon ring, again six-membered but this time two-storied, with the formula $[Si_{12}O_{30}]$. This ring could be considered as a natural doubling of the single-story six-membered ring $[Si_6O_{18}]$ found in beryl, particularly as milarite is itself a beryllium compound, the two formulas being

beryl - (He) $Be_3Al_2[Si_6O_{12}]$
milarite - K $(Be_2Al)Ca_2[Si_{12}O_{30}]$.

2

The Second "Chapter" of Silicate Crystal Chemistry

The strongly-held view that rapid development of silicate crystal chemistry was impossible was also bound up with the purely technical difficulties involved; the simpler examples had all been exhausted, leaving only complex structures involving more than 20 or 30 variables, and it was impossible to solve such structures with the techniques formerly available. A definite breakthrough occurred in the period 1948-1952, when new "direct" methods were worked out, first abroad, and then, perhaps predominantly, here; these allowed a solution even with a large number of unknowns, the only limitation being that the number of atoms in the unit cell should not exceed 80-100. Unlike the classical methods, these new approaches were effective with crystals of low symmetry; in particular with monoclinic symmetry, and the greater part of the crystals met with in mineralogy (the silicates especially) belong to just this symmetry system. It was found possible to pass over to machines the complex calculations involved in structural analysis. Beginning in 1953, the Moscow laboratories, working in close collaboration with the Gorky and Baku collectives, worked out a whole series of silicate structures (ilvaite, $CaFe^{...}Fe_2^{..}Si_2O_8OH$ [5], epidote [6], zoisite [7] $Ca_2Al_2FeSi_3O_{12}OH$, cuspidine $Ca_4[Si_2O_7]F_2$ [8], xonotlite $Ca_6[Si_6O_{17}](OH)_2$ [9], wollastonite $CaSiO_3$ [10], gadolinite $Fe^{..}Y_2Be_2O_2Si_2O_8$ [11], seidozerite $Na_2Mn_{0.5}Ti_{0.75}Zr_{0.75}[Si_2O_8]F$ [12], låvenite $(Na, Ca, Mn)_3(Zr, Fe)Si_2O_8F$ [13], lovozerite $Na_2ZrSi_6O_{15} \cdot 3H_2O$ [14], epididymite $NaBeSi_3O_7OH$ [15], rhodonite $MnSiO_3$ [16], and hillebrandite $2CaO \cdot SiO_2 \cdot H_2O$ [17]. The results obtained allowed a new appraisal to be made of the whole field of silicate crystal chemistry.

In six of the above structures it was found that, although the empirical formula was of the orthosilicate type, the basic silicate radical was the diortho group $[Si_2O_7]$, a group which had been almost wholly neglected during the "classical" period from 1930 to 1953. In all the rest of the above structures, the binuclear group $[Si_2O_7]$ was prominent as the fundamental link in the oxygen-silicon chains, ribbons, networks, and rings discovered, which greatly differed in their geometry from the classical Bragg chains, ribbons, networks, and rings. The two-storied six-membered ring previously found in milarite could be fitted into this series if it was considered as the result of condensing six Si_2O_7 diortho groups, according to the scheme $6[Si_2O_7] - 6 \cdot 2O = [Si_{12}O_{30}]$.

Almost all the above minerals are either pure calcium silicates, or are silicates with Zr, Ti, Nb, Ta, Mn, RE, and other "refractory" elements, with Ca or Na always present. With these large cations present in silicates the structural unit ("Baustein") is the $[Si_2O_7]$ group, instead of the $[SiO_4]$ group which we find in the Bragg silicates, where the cations are mostly the smaller Mg, Fe, and Al ions. In the light of these purely structural results, we have to make the apparently paradoxical inference that the structure of a silicate is governed, not by the silicon atoms, and not by the oxygen-silicon anions, but conversely by the cations, which are usually arranged in columns of oxygen octahedra (around each cation), so that it is these which make up the basic structural framework, and the oxygen-silicon radicals are only adapted to fit them. Silicon, silica, which is extremely unreactive in ordinary chemical analysis, appears just as inert in the formation of crystalline structures. The oxygen-silicon radicals — chains, ribbons, networks, and even rings — are quite strong, but not too rigid, and so they are easily deformed and then adapted to fit the various structural conditions demanded by the arrangement of the cations present.

We stated above that a six-membered oxygen-silicon ring was discovered in dioptase, so that the basic feature of beryl was virtually repeated. This is not quite true, since the beryl ring $[Si_6O_{18}]$ has maximum symmetry (a 6-fold axis and a plane of symmetry), while in dioptase three of the tetrahedra in the ring have their vertices pointing upward and the other three have them pointing downward, in accordance with a rhombohedral lattice. This is connected with the fact that the core, the crystal "nucleus" of the structure of dioptase, $Cu_6[Si_6O_{18}] \cdot 6H_2O$, is a ring of six H_2O particles, shaped as if cut out of the ice structure, with its characteristic rhombohedral symmetry (cf.[18]). This appears to be sufficient to impress rhombohedral symmetry on the six-membered oxygen-silicon ring above and below the $(H_2O)_6$ "ice ring" in dioptase, and subsequently on the whole structural layout of this ring silicate.

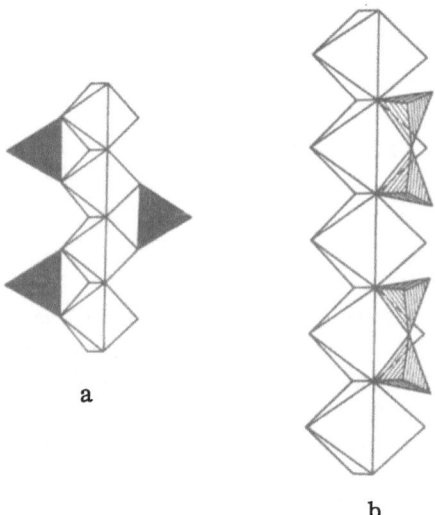

a

b

Fig. 1. The elementary silicate units (building bricks) in silicate crystal structures: (a) the $[SiO_4]$ ortho group in the first "chapter"; (b) the $[Si_2O_7]$ diortho group in the second "chapter."

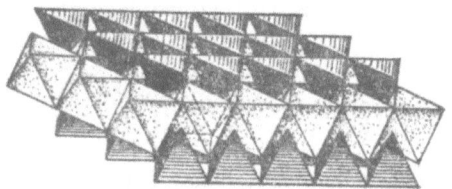

Fig. 2. Close-packing of octahedra, separated by double the number of tetrahedra in two orientations.

Thus classical (Bragg) silicate crystal chemistry, having held the field for 30 years, must now be considered as only the first "chapter" in the subject, concerned with the comparatively "small" cations Mg, Fe, and Al, which form around themselves oxygen octahedra of edge 2.7-2.8 A, i.e., "commensurable" (in Goldschmidt's terms) with the edge of the Si tetrahedron (2.55-2.7 A).

With larger cations (primarily Ca and Na), which have octahedra with edges "incommensurable" with the edges of a Si tetrahedron (3.8 A as against 2.6 A), the basic silicate unit becomes the larger Si_2O_7 group (the edge of the trigonal prism into which this group is inscribed has a length of 4-4.2 A, "commensurable" with the edge of a Ca (or Na) octahedron, as is seen from Fig. 1a, 1b). This is the first representative of the second "chapter" of silicate crystal chemistry [19].

Isolated SiO_4 tetrahedra, and the commensurable Mg octahedra can be comparatively easily arranged into a joint close-packing with the anions O, OH, and F.

As we know [18], in close-packing the number of octahedral holes (octahedra) is equal to the number of anions; there are twice as many tetrahedra, and these are present in two orientations (Fig. 2). Determination of the structure of a silicate crystal built on the close-packing principle therefore boils down to finding out which octahedra and tetrahedra are empty and which are occupied by the various cations [18].

This property of making a joint packing with separate Mg octahedra is still applicable to complex oxygen-silicon radicals made up of "single" SiO_4 tetrahedra, i.e., chains, ribbons, and networks of hexagonal rings belonging to the first "chapter." In this way are formed minerals with high specific gravities (> 3), usually dark in color (melanocratic), because of the high degree of isomorphous replacement of Mg and Al by "colored" cations of small radius, including Ti, V, Cr, Mn, Co, Ni, and above all Fe, in its two oxidation states. Structures of this type include olivite, topaz, the chondrotite-humite series, many pyroxenes and even more amphiboles, micas, chlorites, and clay minerals (these are all important members of the first "chapter").

With large cations, where the basic silicate unit is the Si_2O_7 group, the open radicals based on this (their characteristic feature is the eight-membered ring) form light rock-forming minerals with low specific gravities (< 3); they are light-colored (leucocratic) because all the principal large cations, K, Na, Ca, RE, are "colorless," since they have a complete octet in the outer electron shell.

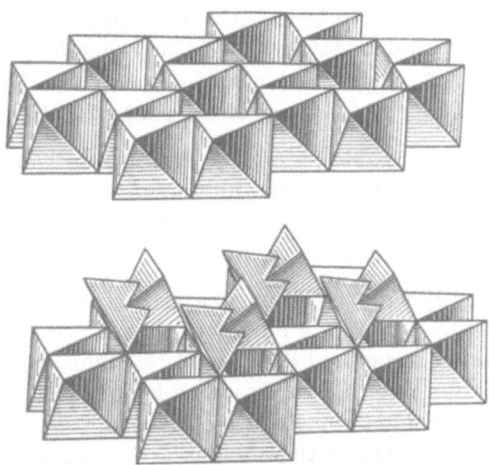

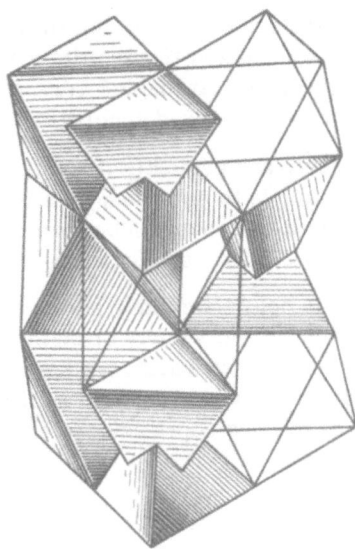

Fig. 3. Structure of thortveitite, $Sc_2Si_2O_7$ (and also of $Mg_2P_2O_7$). In this two-storied hexagonal packing, a layer of Sc octahedra with a corundum-muscovite motif alternates with a layer of Si_2O_7 diortho groups, inscribed in octahedra, with a carbonate motif.

Fig. 4. The structure of Ba spinel, $BaAl_2O_4$, containing large octahedra of three sorts: one is occupied by a Ba atom, the second by an Al_2O_7 group, and the third is empty.

The beryl-milarite pair discussed above serves as an interesting illustration of this statement. From their structures and formulas it follows that in beryl the central cation in the octahedra is Al, while in milarite it is Ca (in milarite Al only acts as partial replacement for Be in the tetrahedra). The aluminum core matches with ortho groups, SiO_4, while the calcium core needs diortho groups, Si_2O_7. Both typical units are joined up into six-membered rings, single-storied in milarite. Beryl (emerald) is a dark-colored mineral, milarite is light-colored.

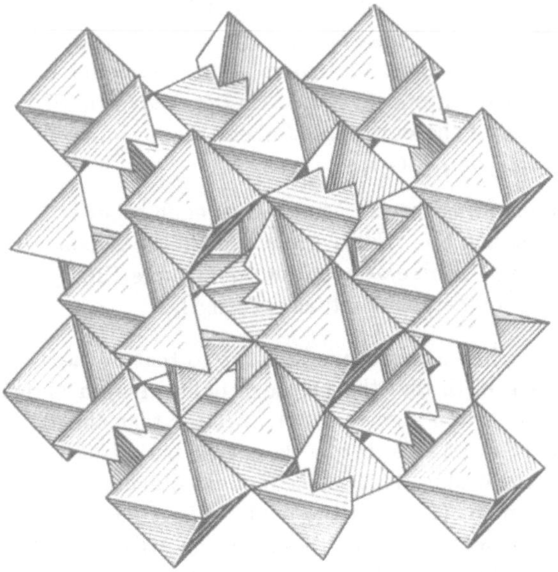

The joining-up of two tetrahedra into Si_2O_7 groups may be centrosymmetric; the centrosymmetric group formed can be inscribed in a somewhat distorted octahedron (a twisted trigonal prism) which, in close packing with other octahedra (of Ca, Na, RE), forms the structure of the only diortho silicate given in Bragg's book (also in the majority of books on crystal chemistry, or in the crystal chemistry sections of mineralogy courses), the rare mineral thortveitite, $Sc_2[Si_2O_7]$.

Fig. 5. The cubic structure ZrP_2O_7, consisting of Zr octahedra and octahedra containing P_2O_7 diortho groups.

Fig. 6. The structural component of Ca(Na) diorthosilicates —the tilleyite ribbon. The trigonal prisms containing the Si_2O_7 groups are crammed into the "troughs" of the "waves" formed by the columns of Ca octahedra.

In Fig. 3, the reader who is familiar with the terminology used in the author's "blue book" [18] will see how in the idealized thortveitite structure a "corundum-muscovite" layer of Sc octahedra alternates with a layer of "carbonate" arrangement of octahedra, each containing diortho groups [Si_2O_7]. Two kinds of layers are packed according to the rule of two-layer hexagonal close-packing [18].*

A more usual, and evidently (from what follows) more favorable, way of joining two tetrahedra into a diortho group, is to join them by a (pseudo) plane of symmetry instead of a center, and in this case the resulting Si_2O_7 group can be inscribed in a trigonal prism. We can also describe this as a prism formed by six O atoms, containing a binuclear $Si-O-Si$ cation.

The prism is not particularly suitable for packing jointly with the octahedra around large cations, and so there is not a great variety of methods of combining the two. In a series of Ca and Na silicates, in particular the cement silicates, we have the "ready-made" structural component or "increment" shown in Fig. 6, which we have called a tilleyite ribbon, after the mineral in which it was first detected. In Fig. 6 the rhombs represent the end octahedra of infinite columns of Ca octahedra resting on their edges. The columns themselves are perpendicular to the plane of the drawing. The columns are joined together by means of the side edges of their octahedra into a corrugated sheet ("wall") stretching in the bc plane. The black triangles represent Si_2O_7 groups inscribed in trigonal prisms and crammed between the Ca octahedra in such a way that, along the direction of the columns, a layer of octahedra with crammed Si_2O_7 groups alternates with a layer of octahedra without Si_2O_7 groups. In the adjacent half-wave of the corrugated sheet the Si_2O_7 group may be on the same level as in the first, or may be on an alternate level (in different structures).

In the structure of tilleyite itself [20], the ribbons are discrete, and interconnected by extra Ca octahedra (in columns) and CO_3 groups, as shown diagrammatically in Fig. 7. The repeat distance along the a axis, perpendicular to the plane of the sheet, includes two tilleyite ribbons, and two connecting layers of Ca octahedra plus CO_3 groups, making six layers of Ca octahedra in all. In each half-wave of the corrugated sheet there is one formula unit, $Ca_5[SiO_7](CO_3)_2$.

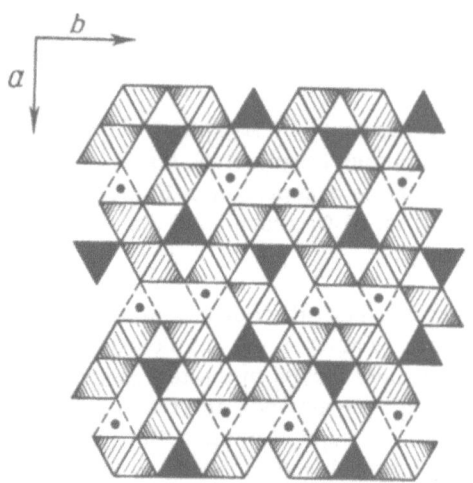

Fig. 7. Structure of tilleyite, $Ca_5[Si_2O_7](CO_3)_2$

In the structure of cuspidine [8], the tilleyite ribbons are directly attached to each other. There are two possible methods of connection, shown in Figs. 8a and 8b. Nature has chosen the second alternative, in which the double links of the tilleyite ribbon are each extended into quadruple links, which are interconnected in a pattern which gives the impression of great structural strength.

*The magnesium phosphate familiar to analysts, $Mg_2P_2O_7$, has a structure similar to that of thortveitite. The ionic radius of P is so much smaller than that of Si that the Mg here behaves as a "large" cation, requiring around itself a bigger P_2O_7 radical. The Al ion also is much smaller than the Ba cation, and in the structure of "barium spinel," $BaAl_2O_4$ (Fig. 4), the large Ba octahedra alternates with similar large octahedra containing [Al_2O_7] groups. Here also we have hexagonal close-packing. Each of the two alternating layers has an ilmenite arrangement [18], but the packing is such that along each 3-fold axis there extends a continuous column of Ba octahedra. Fig. 5 shows the structure of ZrP_2O_7, in which Zr octahedra alternate with octahedra containing P_2O_7 groups in a ReO_3 type structure. The overall cubic symmetry is retained on the pyrite principle; the axes of P_2O_7 groups are stretched along four nonintersecting 3-fold axes of the cubic cell.

In this set of drawings the Si_2O_7 (P_2O_7) group is shown in the form of two intersecting tetrahedra. This has been done so that it can be easily stacked using ordinary (regular) octahedra, instead of distorted ones, as they would have to conform with an elongated Si_2O_7 (P_2O_7) group.

If we examine the formula of cuspidine, $Ca_4[Si_2O_7]F_2$, it is easy to see how this corresponds to the formula of the tilleyite ribbon if, instead of introducing three oxygen layers in each "increment," we only introduce two (or more accurately, one and two halves, which on condensation will give a single layer shared between two ribbons). For every two oxygen layers there are 4×2 atoms of O(F). The ninth O atom, the central one of the Si_2O_7 group, is not taken into account.

a

In tilleyite the two F(OH) atoms in cuspidine are replaced by two O atoms from the two CO_3 groups. The other two pairs of O atoms from the CO_3 groups form a set of four O atoms for the third layer, in which the fifth Ca atom is placed (in an octahedron). Thus the formula developed for tilleyite, after separating the cuspidine "molecule" (tilleyite ribbon), must be written as $Ca_4[Si_2O_7]Ca(OCO_2)_2$.

b

Fig. 8. Two modifications of the structure of $Ca_4[Si_2O_7]F_2$: (a) with double Ca octahedron links; (b) with quadruple links (cuspidine structure).

In the structure [21] of one of the most important cement silicates, tricalcium silicate hydrate (TSH in cement terminology), we again have tilleyite ribbons (or "hydroxy cuspidine molecules") interconnected by a third layer, in this case with two "molecules" of $Ca(OH)_2$, the four hydroxyl groups of which provide the third oxygen layer, and the two Ca atoms, the third layer of octahedra (Fig. 9). Thus the formula of TSH can be written $6CaO \cdot 2SiO_2 \cdot 3H_2O = Ca_4[Si_2O_7](OH)_2 \cdot 2Ca(OH)_2 = Ca_6[Si_2O_7](OH)_6$. In the second formula the tilleyite ribbon (cuspidine "molecule") has been written separately, with the $Ca(OH)_2$ groups relegated to a "side chain." The first formula is the classical one, in oxide terms, with a ratio of $Ca : Si = 3 : 1$.

These structures are given to illustrate the fact that the role played by the single SiO_4 tetrahedron, in structures principally consisting of $Mg(Fe^{..})$ and $Al(Fe^{...})$ octahedra, is taken over by the diortho group, Si_2O_7, in structures made up of Ca(Na, RE) octahedra.*

The necessity to distinguish between large and small cations in silicate formulas, at least where structures were concerned, was appreciated in the early years of silicate crystal chemistry, when formulas were written in the form $A_mB_n[(Si, P)_p(O, OH, F)_q]$, where A stood for large cations such as Ca, Na, K, RE, and B for small ones such as Mg(Fe), Al, Ti.

---

*This is an extremely important condition. If the Ca(Na) does not fill an octahedron with 12 edges, but, for example, is a twisted cube with 16(18) edges, then two of these edges may be shortened to be commensurable with the edges of a SiO_4 tetrahedron; the characteristic feature of the structure of garnet [18] is the presence of chains of alternating Ca polyhedra and single SiO_4 tetrahedra. The same thing is found when we have "normal" polyhedra [18] for the coordination number CN 8, in the structures of zircon, $ZrSiO_4$, xenotime, YPO_4, and anhydrite, $CaSO_4$, and also in gypsum, $CaSO_4 \cdot 2H_2O$ [18]. One of the simplest ways of adapting a Ca octahedron for single Si tetrahedra is to exchange one half of the octahedron for a trigonal prism, forming a polyhedron of CN 7, in which one edge of the prism can be somewhat shorter than the others, so that it can be coupled to the edge of a Si tetrahedron (as in the structure of sphene, $CaTiSiO_5$, for example). A Ca polyhedron with CN 8 is coupled in roughly the same way in the structure of diopside.

We have outlined above the fundamental differences between the oxygen-silicon radicals in pure "A" silicates and pure "B" silicates, for the cases where A and B have CN 6, i.e., they lie inside oxygen octahedra (of different size).

For many cations, however, no indication can be given beforehand as to whether they will behave as "A" or "B" in an unknown structure. The best-known example is the Mn^{2+} cation, which on interpretation of the chemical analysis is usually included partly in the A group, with Ca, Na, and RE and partly in the B group, with $Fe^{..}$ (Mg).

In the structure of cuspidine, $Ca_4[Si_2O_7]F_2$, (in tilleyite ribbons) there is no doubt that all the Ca ions behave as "A" cations only. However, when we have a column (see Fig. 1b) of octahedra which alternate between "bound" (to Si_2O_7 groups along their length, and therefore "stretched") and "free" (not connected to Si_2O_7 groups), it is possible to populate the latter quite freely with "ambiguous" cations such as Mn^{2+}, Zr, and even Ti. Minerals of the woehlerite-låvenite group have this structure, and as Table 1 shows they are all actually variations on the pure Ca structure cuspidine, so from now on they should be known as members of the cuspidine-woehlerite-låvenite group.

Figure 10 shows the result of a detailed analysis of the structure of låvenite, according to V. I. Simonov [13]. The geometrical plan of låvenite (Fig. 10) is exactly the same as that of cuspidine (Fig. 8b); however, while in the latter all four columns of octahedra are

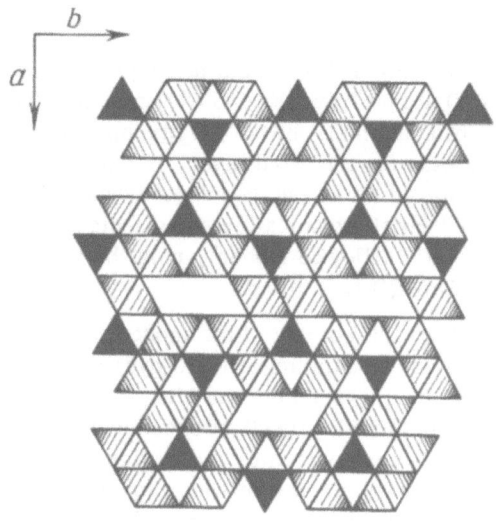

Fig. 9. Structure of tricalcium silicate hydrate, $TSH = Ca_4[Si_2O_7](OH)_2 \cdot 2Ca(OH)_2 = Ca_6[Si_2O_7](OH)_6$.

TABLE 1. Minerals of the woehlerite-låvenite group

Mineral and formula	a	b	c	α	β	γ	z
Woehlerite - $Ca_2NaZR[Si_2O_7]OF$	10.82	10.28	7.27	–	108°57'	–	4
Låvenite - $(Na,Ca,Mn)_3(Zr,Ti,Fe)[Si_2O_7]OF$	10.95	10.01	7.19	–	110°18'	–	4
Hellandite - $(Ca,RE,Mn)_3(Al,Fe)[Si_2O_7]O \cdot H_2O$					109°45'		
Guarinite - $(Na,Ca)_3(Fe,Ti,Zr)[Si_2O_7]OF$	10.93	10.31	7.33	90°29'	108°50'	90°08'	4
Rosenbuschite - * $(Na,Ca)_3(Fe,Ti,Zr)[Si_2O_7]OF$	10.14	11.41	7.28	91°21'	99°38'	111°54'	4
Hiortdahlite - $NaCa_2Zr[Si_2O_7]OF$	10.91	10.29	7.32	90°29'	108°50'	90°08'	4
Niocalite - $Ca_{3.5}Nb_{0.5}[Si_2O_7]O(OH,F)$	10.83	10.42	7.38	–	109°40'	–	4
Cuspidine - $Ca_4[Si_2O_7](F,OH)_2$	10.35	10.43	7.55	–	110°04'	–	4

* As a result of our recent analysis, rosenbuschite passed into the group of rinkite (R.P. Shibaeva, N. V. Belov, Doklady AN SSSR, 143, 1428, 1962).

occupied by Ca atoms, in låvenite, in agreement with the formula $CaNaZr(Fe,Mn)[Si_2O_7]OF$, we have two columns in which large, "light" Ca octahedrons alternate with "colored" Zr octahedra, and two in which "light" Na octahedra alternate with "colored" Fe(Mn) octahedra (see Fig. 10b, a front view).

There is another mineralogical class with the same formula as the cuspidine-woehlerite-låvenite group, but with different lattice parameters. This is the rinkite-lamprophyllite group (Table 2). At first sight the main difference between the two groups of minerals is that the first contains Zr, which is almost absent in the second group. So when E. I. Semenov discovered the new mineral seidozerite [22] in 1957, with the same type of formula, although with Ca completely replaced by Na, and, more important, with a high Zr content, he immediately placed it in the woehlerite group. A detailed X-ray structural analysis [12] very soon placed seidozerite at the head of the second group, just as it had placed cuspidine at the head of the woehlerite-låvenite group, and explained the reasons, not only for the differences, but also for the great structural similarities between the two groups of minerals.

We have shown above two methods of coupling tilleyite ribbons together, in which the octahedron links look the same in two directions. But it is also possible to couple the ribbons together in such a way that the octahedron links continue in one direction to infinity, while in the other direction they remain only double (Fig. 11a). In the seidozerite structure, this arrangement is modified so that for each jump along the b-axis, over the "infinite" link, the "double" links are displaced along the a-axis by one story (Fig. 11b).

If in cuspidine, tilleyite, and TSH there is an inherent (pseudo) orthorhombic arrangement, with mutually perpendicular a and b axes, then seidozerite should show a monoclinic arrangement, with a doubled a lattice parameter (because of the jump) and a halved b lattice parameter. This outwardly expressed monoclinic character (with the originally supposed chemical differences) forces us to assume that the two groups are different in kind, with cuspidine at the head of one and seidozerite at the head of the other.

We can see now that both groups of minerals with the formula $(Ca, Na, \ldots)_4[Si_2O_7](O, OH, F)_2$ are extremely close, and they should be spoken of not as polymorphs (in the wider sense of the word, i.e., polymorphic structural types) but only as polytypes.

The structure of seidozerite, $Na_8(Zr_3Ti)Ti_2Mn_2O_4[Si_2O_7]_4F_4$, has been analyzed in particular detail, and the three projections shown in Fig. 12 demonstrate very clearly how, in each column, a lengthened octahedron around Na alternates with a shortened octahedron around either Zr (in two columns), Mn (in one column) or Ti (in one column), while each $[Si_2O_7]$ group is connected on all three sides with Na octahedra only.

While cuspidine, $Ca_4[Si_2O_7]F_2$, which we placed at the head of the woehlerite-låvenite group, clearly showed the presence of a Si_2O_7 group (including in this all the oxygen in the mineral) directly

TABLE 2. Minerals of the rinkite group

Mineral and formula	a	b	c	α	β	γ	z
Rinkite - $Na(Ca,Ce)_2(Ti,Ce)[Si_2O_7]OF$	18.51	5.68	7.47	–	91°13'	–	4 •
Johnstrupite - $(Ca,Na)_3(Ti,Ce)[Si_2O_7]OF$					93°04'	–	4
Mosandrite - $(Ca,Na,RE)_{2-3}(Ti,Zr,Ce)[Si_2O_7]O(OH,F)$	18.41	5.64	7.43	–	93°	–	4
Lamprophyllite - $Na_3Sr_2Ti_3\{[Si_2O_7]O(O,OH,F)\}_2$	19.09	5.35	7.06	–	–	–	2
Seidozerite - $Na_2Mn_{0.5}Zr_{0.75}Ti_{0.75}[Si_2O_7]OF$	18.30	5.53	7.10	–	102°43'	–	4

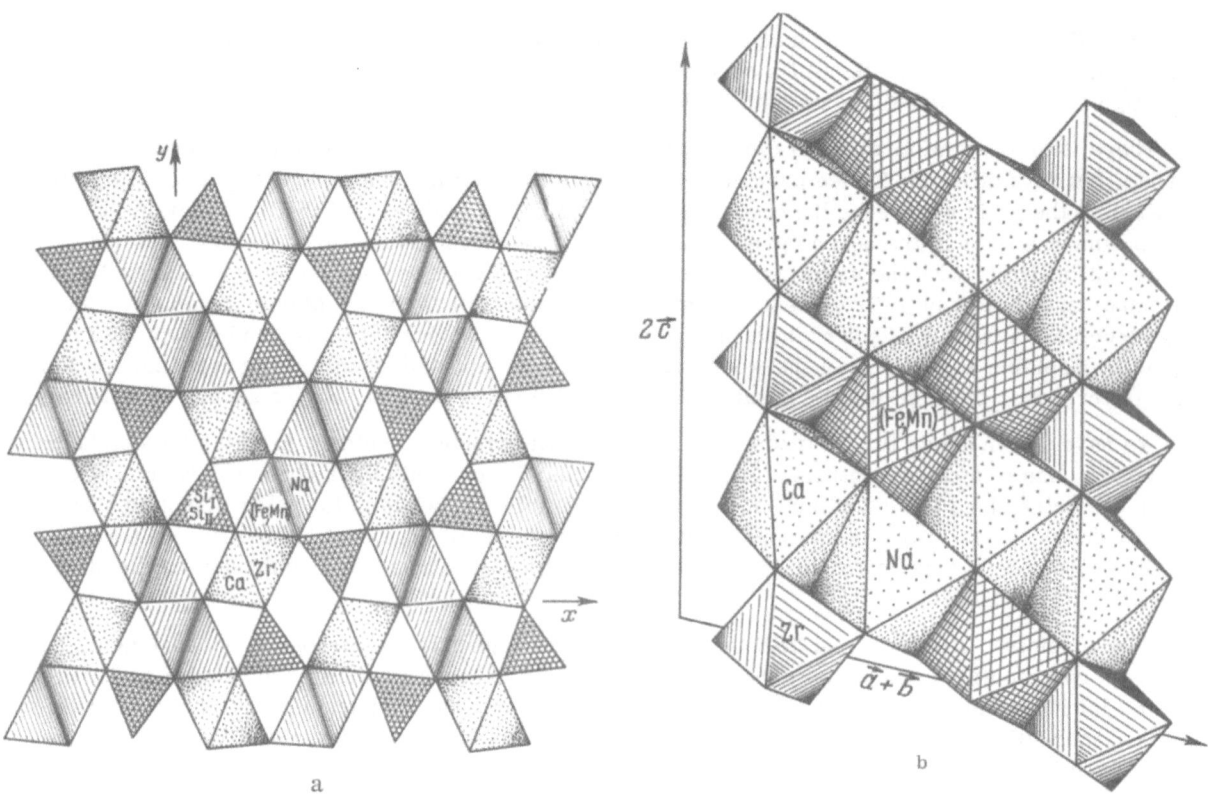

Fig. 10. The structure of låvenite, $CaNaZr(Fe, Mn)[Si_2O_7]OF$: (a) plan; (b) front view. In one pair of columns Ca and Zr octahedra alternate, in the other Na and (Fe, Mn) octahedra alternate.

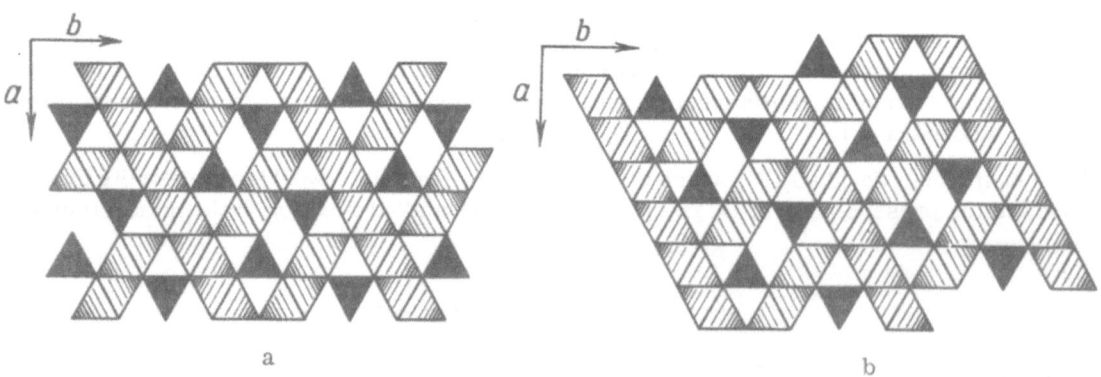

Fig. 11. Structure of seidozerite, $Na_4Zr_2TiMnO_2[Si_2O_7]_2F_2$: (a) ideal plan; (b) actual plan.

from its chemical analysis, seidozerite was initially written in the orthosilicate form as $Na_8Mn_2Zr_3Ti_3[SiO_4]_8F_4$, also from its chemical analysis, and it was only as a result of the X-ray structural analysis that the eighth O atom was taken out of the square brackets (containing the silicate radical) and placed with F, to play the same part as the second F(OH) atom in cuspidine.

A similar chemical-analytical operation was performed on some "classical" orthosilicates, ilvaite and epidote-zoisite [5, 6, 7], when we again helped split up an "orthosilicate" into Si_2O_7 groups and "free" oxygen (with the aid of the new chemical weapon—X-ray structural analysis), the oxygen being relegated to the same functional classification as the F(OH), giving

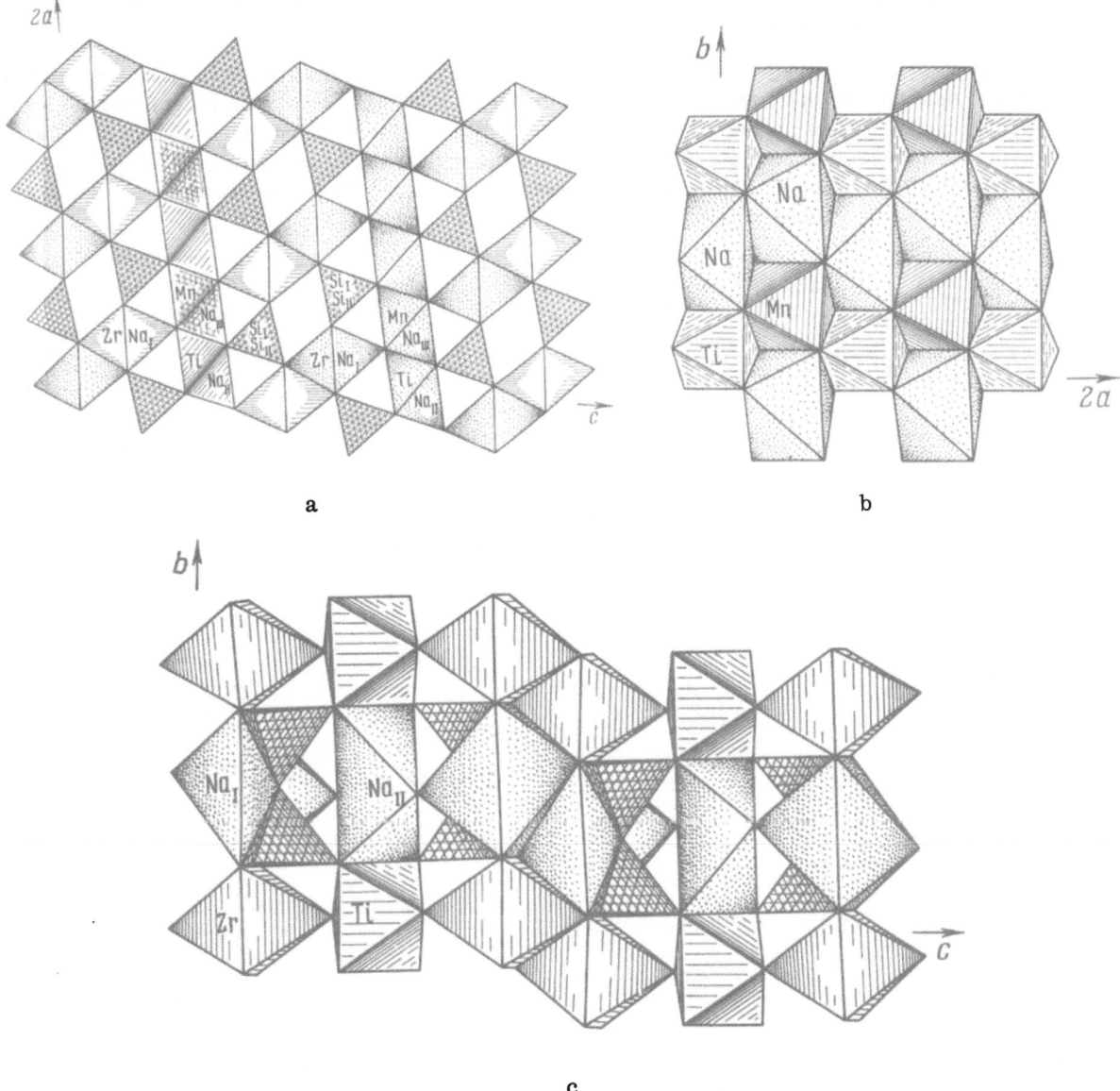

Fig. 12. Alternation of Na octahedra with three types of "colored" octahedra, Zr, Ti, and Mn, in the structure of seidozerite: (a) plan (projection along the octahedron column axis); (b) front view of "wall" of octahedra formed around various cations; (c) [Si₂O₇] groups attached only to elongated Na octahedra.

$$\text{ilvaite} - \text{CaFe}^{\cdots}\text{Fe}_2^{\cdot\cdot}\text{Si}_2\text{O}_8\text{OH} = \text{CaFe}^{\cdots}\text{Fe}_2^{\cdot\cdot}\text{O}[\text{Si}_2\text{O}_7](\text{OH})$$
$$\text{epidote-zoisite} - \text{Ca}_2\text{Al}_2\text{FeSi}_3\text{O}_{12}\text{OH} = \text{Ca}_2\text{Al}_2\text{FeO}[\text{SiO}_4][\text{Si}_2\text{O}_7](\text{OH})$$

In the orthosilicate-type formula of the first mineral, every eighth O atom is not connected to Si, and in the epidote formula every twelfth O atom is not connected. These "free" O atoms in both minerals play a similar role to the OH groups.

11

In epidote-zoisite both A cations (Ca) and B cations (Al, Fe) are quite strongly represented, and it is curious to see that the Si_2O_7 group is connected along its length with the edge of a Ca octahedron, while it is connected with Al octahedra by the edges of its base.

On the other hand, in ilvaite, with a predominance of B cations (Fe), the Si_2O_7 group is not connected with the side of a Ca octahedron. Both its tetrahedra are joined only with Fe octahedra, and the Si_2O_7 group remains (as emphasized in the detailed paper on ilvaite [5]) only as a relic of a pyroxene chain, in a structure intermediate between those of the orthosilicate andradite, $Ca_3Fe_2[SiO_4]_3$, and the metasilicate hedenbergite, $CaFe[Si_2O_6]$. In connection with this, the $Si-O-Si$ angle in ilvaite is only about 120°, while in epidote-zoisite it is about 158°.

Isomorphism Relations Between Zirconium and Titanium

In the Fersman "isomorphism stars" for both zirconium and titanium the two do not fall in line, which is natural in view of the great difference between the Goldschmidt ionic radii: $Zr-0.87\text{Å}$, $Ti-0.64$ Å. The difference of 36% (calculated on the smaller radius, according to Goldschmidt and Fersman) is much greater than the 15% which is usually quoted as the limit for isomorphism. However, ever since the Lovozero minerals were brought to light, more and more has been said about isomorphism between Zr and Ti [23].

Replacement of Zr by Ti has been clearly demonstrated in astrophyllite-kupletskite, and again in låvenite (Ti-låvenite), but a particularly clear example is a mineral dealt with in the preceding section, seidozerite. The detailed and reliable results of an X-ray analysis (as yet not carried out for kupletskite) showed that in seidozerite, almost exactly a quarter of the Zr atoms were replaced by Ti, while two thirds of the Ti atoms occupied their own positions, in which there were no Zr impurity atoms.

An examination of the chemical analyses provides a simple reason for this isomorphism. The three minerals contain appreciable quantities of manganese, which characteristically appears in different låvenites [22]. At 6% MnO, the ZrO_2 and TiO_2 contents are 23.20 and 5.28%, respectively, and at 10.34% MnO they are 16.72 and 11.30%. In astrophyllite the Zr : Ti ratio varies very widely, with a low overall content (~14%), but the end Mn member, kupletskite, with 27% MnO, contains only TiO_2, which here has completely replaced ZrO_2.

The isomorphous relationships in seidozerite are curious. We gave its formula above in terms of Mn, but in this, E. I. Semenov included all the Mg found. However, the ratio of MnO : MgO in this analysis was 1 : 1, so we can also say that here Mn is showing isomorphous replacement of Mg (in spite of the difference of 16% between the ionic radii, $Mg^{\cdot\cdot}-0.78$, $Mn^{\cdot\cdot}-0.91$ Å).

We can arrive at a solution to this paradox if we assume that trivalent Ti is not so very exceptional, as was shown by A. I. Tsvetkov [24] in his work on synthetic pyroxenes. If we write down the equation

$$Ti^{4+} + Mn^{2+} \rightleftarrows Ti^{3+} + Mn^{3+} \ (\tfrac{1}{2} \ Mn^{4+} \ ?),$$

we can see that the values of ionic radii for the cations on the right-hand side ($Ti^{3+}-0.83$, $Mn^{3+}-0.71\text{Å}$), are such that there remains no obstacle to isomorphous replacement, both by part of the Ti in the Zr octahedra, and by all the Mn in the Mg octahedra.

It has frequently been pointed out that a very minor influence (even just a tendency) is often sufficient to determine which of two or three possible forms will be taken by an element, or even a compound (e.g., a small quantity of SrO stabilizes $CaCO_3$ in the form of aragonite). Such an influence appears to have a decisive effect in the isomorphous replacement of zirconium by titanium (but not vice versa) in the three minerals containing appreciable amounts of Mn mentioned above.

It should be noted that the Ti^{3+} pyroxenes prepared by A. I. Tsvetkov showed strong pleochroism, from dark red to red and yellow. It may be coincidence, but exactly the same thing was very clearly observed in the seidozerite crystals [22].

Oxygen-Silicon Chains and Ribbons in the Second "Chapter" of Crystal Chemistry of Silicates

Proceeding along the same lines as for the first "chapter," we will now pass on from "single" $[Si_2O_7]$ groups, first to the analogs of two types of metasilicate radicals, chains closed into rings, and infinite chains.

Examples of Ca silicates with Si_2O_7 groups condensed into two-storied, six-membered rings $[Si_{12}O_{30}](= 6Si_2O_7 - 6 \cdot 2 O)$, have already been mentioned; these are milarite, and high cordierite-osumilite. It appears that elpidite, $Na_2Zr[Si_6O_{15}]$, [25-26], has a two-storied, three-membered ring $[Si_6O_{15}] = 3Si_2O_7 - 3 \cdot 2 O.*$

a   b

Fig. 13. Development ("genesis") of pyroxenoid chain, $[SiO_3]_\infty$ $= [Si_{2+1}O_9]_\infty$: (a) Si_2O_7 group of greater height than edge of Ca octahedron; (b) extreme shortening by SiO_4 group of distance between two free corners of adjacent Ca octahedra in the same column.

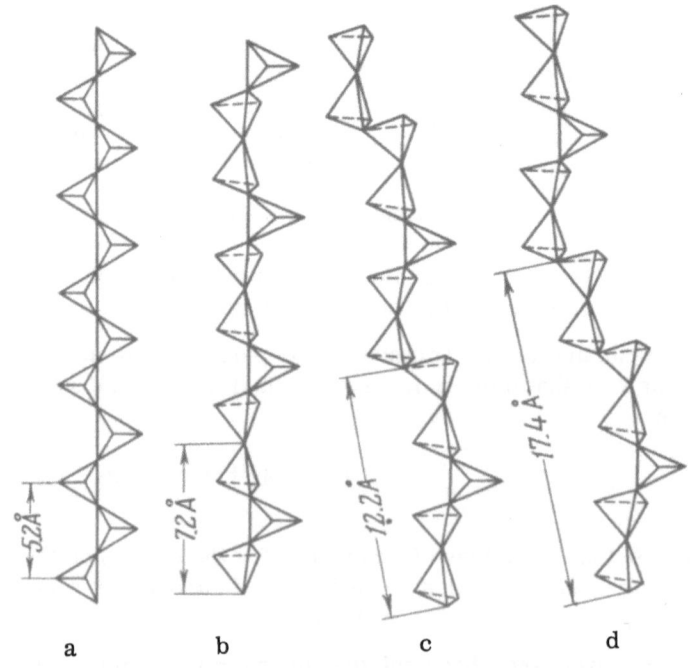

a   b   c   d

Fig. 14. Different types of metasilicate chains with formula $[SiO_3]_\infty$: (a) pyroxene type, link $[Si_2O_6]$; (b) pyroxenoid type, $[Si_3O_9]$; (c) rhodonite type, $[Si_5O_{15}]$; (d) pyroxmangite type, $[Si_7O_{21}]$.

---

*It is not correct. We have shown that very characteristic for elpidite is a ribbon $[Si_6O_{15}]_\infty$ treated later as "epididymite" ribbon (N. N. Nervonova, N. V. Belov, Doklady AN SSSR, 150, 3, 1963).

Both these rings, and also the highly probable two-storied ring $[Si_8O_{20}]$, can be written with the general formula $[Si_2O_5]n$, where $n = 3, 4, 6$.

On moving on to infinite chains, the analogs of pyroxene chains, we meet the pyroxenoid wollastonite [10]. The skeleton in this structure is an infinite column of Ca octahedra laid on edge. As in cuspidine, every second octahedron in the column is coupled to both tetrahedra of a $[Si_2O_7]$ group. Since, however, the height of a SiO_2 group, equal to $4.1-4.2$ Å, is greater than the edge of a Ca octahedron (3.7 Å), the $[Si_2O_7]$ group is deformed, and becomes shortened on the side of the Ca octahedron. The opposite edge of the prism containing the Si_2O_7 group is correspondingly lengthened (Fig. 13a), and to such an extent that it is possible to bridge the gap between two consecutive Si_2O_7 groups on the column with just a single Si tetrahedron, as shown in Fig. 13a.

Thus one should not be too dogmatic in contrasting the silicate radicals of the second "chapter" with those of the first. In each repeat distance along the Ca column of wollastonite we have two Ca octahedra and three Si tetrahedra, one diortho group and one ortho group.* Fig. 13b shows another way of developing a pyroxenoid chain; in this case the silicate chain is not attached to the edges of the Ca octahedra, but to individual corners. The SiO_4 tetrahedron between the "free" oxygen corners of two consecutive Ca octahedra pulls these oxygen atoms together to such an extent that the following pair of oxygen corners can only be bridged by the longer Si_2O_7 group (in wollastonite itself the oxygen-silicon chains are joined to adjacent columns of Ca octahedra by both methods).

If we represent the link distance of the pyroxene chain by two Si tetrahedra, one on one side of the chain axis and one on the other (Fig. 14a), then the link distance of the pyroxenoid chain is represented by three tetrahedra, two on one side and one on the other (Fig. 14b). As the general formula $[SiO_3]\infty$, is the same for both pyroxene and pyroxenoid (wollastonite chains, we can distinguish between them by writing one as $[Si_2O_6]\infty$, and the other as $[Si_3O_9]\infty$, or even $[Si_{1+1}O_6]\infty$ and $[Si_{2+1}O_9]\infty$.[†] In "B"

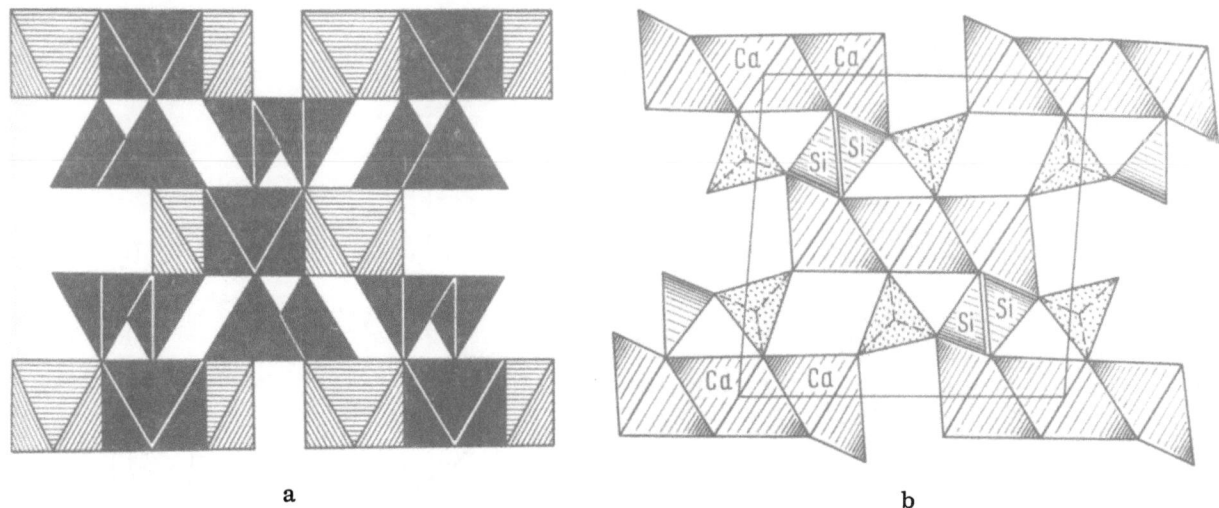

| a | b |

Fig. 15. Basal ("on end") sections through the structures of the principal chain metasilicates. On the odd levels are ribbons of octahedra; the chains of Si tetrahedra are on the even levels: (a) pyroxene, $MgSiO_3$; (b) pyroxenoid, $CaSiO_3$.

---

*The "true" undeniable chain of the second "chapter" put together solely with Si_2O_7 groups has been established in a new silicate batisite, $Na_2BaTi[Si_4O_{12}]\infty O_2$ (A. V. Nikitin, N. V. Belov, Doklady AN SSSR, 146, No. 6, 1962).
† In our new inventory of silicate chains (VI-th Congress IUC-Theses 1963) we give 77 different kinds of chains with the formula $[SiO_3]\infty$.

silicates the general formula for chains closed into rings and for infinite chains is the same, $[SiO_3]_x$, while in "A" silicates the formula for ring radicals is $[Si_2O_5]_x$, and that for chain radicals (single chains) is $[SiO_3]_\infty$.

The sections of both chain structures (pyroxenes and pyroxenoids, Fig. 15a,b) viewed on end are similar. In them we see four levels (stories) of polyhedra (and as many again of layers of O atoms), in a centrosymmetrical arrangement. The odd levels (1, 3, and 5) consist of columns (ribbons) of octahedra, giving a thickness of three octahedra, and the even levels consist of Si tetrahedra. When viewed along the chain axis both structures show two Si tetrahedra per chain, but in the case of pyroxene (Fig. 15a) there actually are two tetrahedra per link, while in the pyroxenoid wollastonite one of the tetrahedra conceals (behind it) a further tetrahedron belonging to the same link. The difference is most clearly seen in a projection perpendicular to the plane of the layers (Fig. 16a,b); we here see not only the different numbers of tetrahedra in the links, but also the different arrangements of the basic skeletons of the structures, the zigzag chains of Mg octahedra in the pyroxenes (Fig. 16a) and the straight columns of Ca octahedra in wollastonite (Fig. 16b).

Treating the matter purely geometrically, we can form an amphibole ribbon from a pyroxene chain by applying a plane of symmetry (Fig. 17c,d). The corresponding ribbon derived by doubling a wollastonite chain is normally formed by application of a horizontal 2-fold axis. (This does not produce a very different effect from a mirror plane in the plane of the projection, Fig. 17e,f.) This rib-

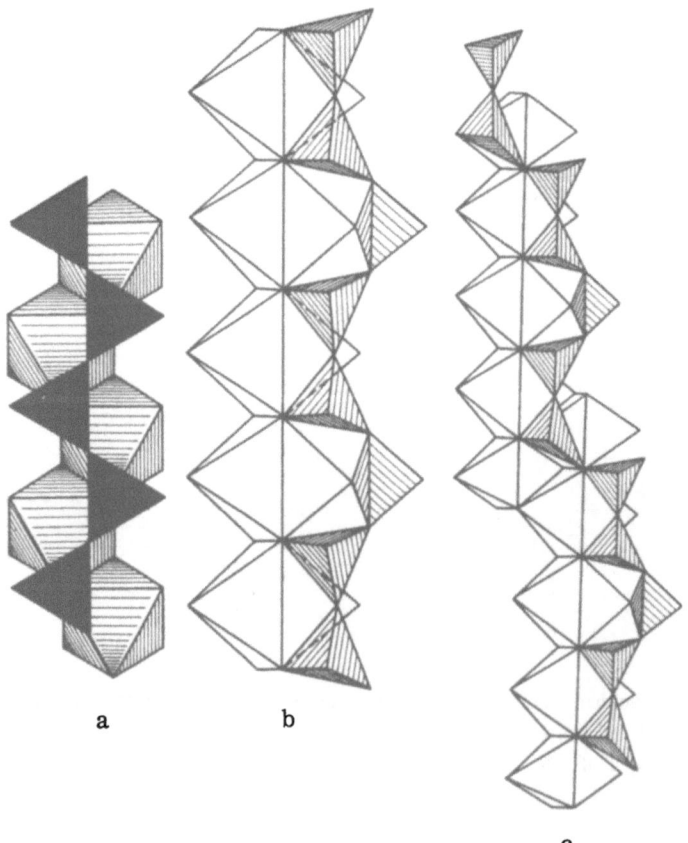

a   b

c

Fig. 16. Correlation between the shapes of the various $[SiO_3]_\infty$ metasilicate chains and the arrangements of the corresponding ribbons of octahedra around principal cations: (a) in pyroxenes; (b) in pyroxenoids; (c) in rhodonite.

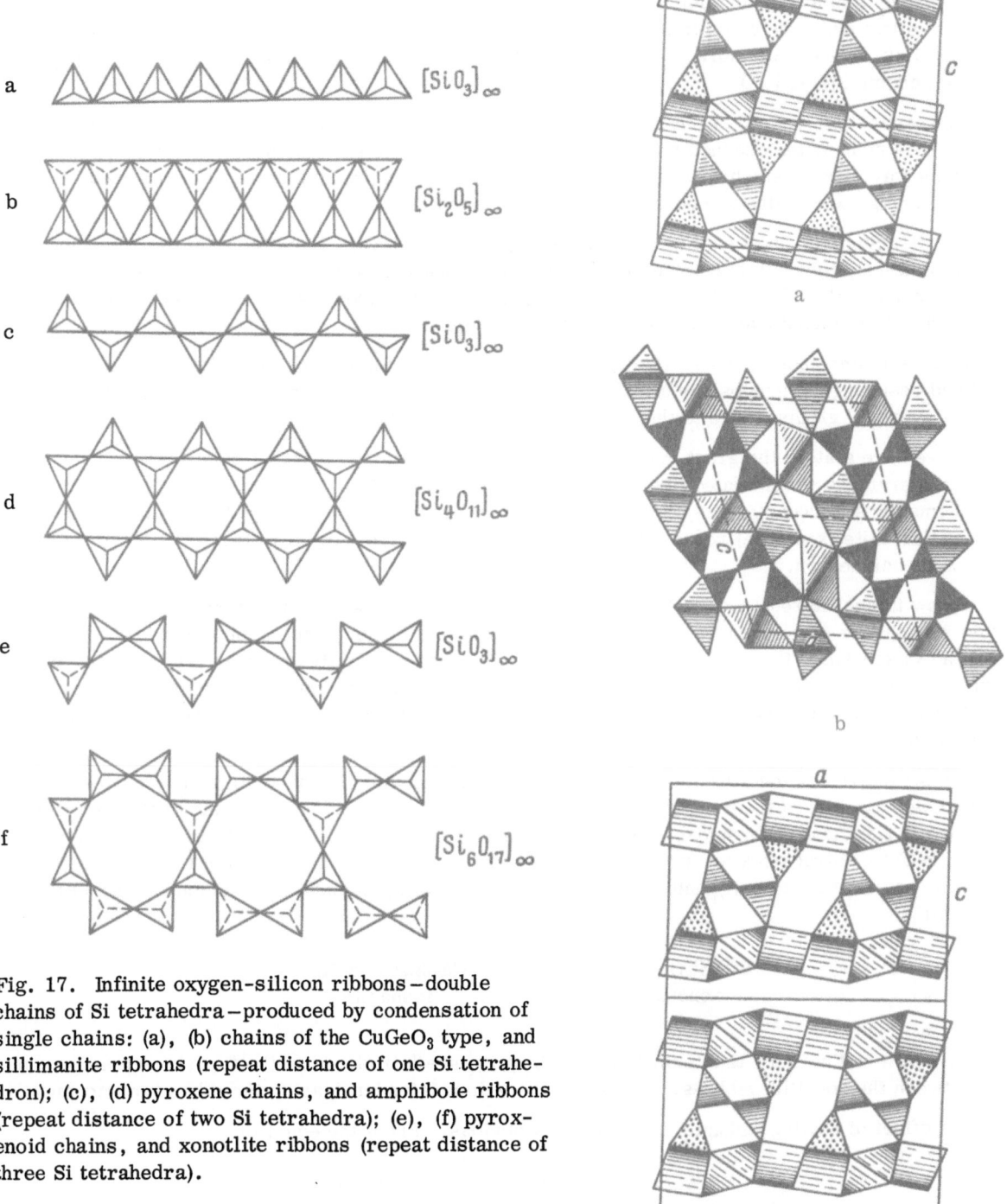

a $[SiO_3]_\infty$

b $[Si_2O_5]_\infty$

c $[SiO_3]_\infty$

d $[Si_4O_{11}]_\infty$

e $[SiO_3]_\infty$

f $[Si_6O_{17}]_\infty$

Fig. 17. Infinite oxygen-silicon ribbons – double chains of Si tetrahedra – produced by condensation of single chains: (a), (b) chains of the $CuGeO_3$ type, and sillimanite ribbons (repeat distance of one Si tetrahedron); (c), (d) pyroxene chains, and amphibole ribbons (repeat distance of two Si tetrahedra); (e), (f) pyroxenoid chains, and xonotlite ribbons (repeat distance of three Si tetrahedra).

a

b

c

Fig. 18. Silicates with xonotlite ribbons: (a) xonotlite; (b) foshagite; (c) hillebrandite.

17

bon is called a xonotlite ribbon, after the mineral in which it was discovered [9]. Since the symmetry element used in the combination process passes through one of the oxygen atoms, the corresponding chemical process is a condensation and will result in the formation of one amphibole ribbon link, $[Si_4O_{11}]$ from two pyroxene chain links $[Si_2O_6]$, according to the scheme $2[Si_2O_6]-O$, and of one xonotlite ribbon link, $[Si_6O_{17}]$ (or more exactly $[Si_{4+2}O_{17}]$) from two wollastonite chain links, $[Si_3O_9]$, according to the scheme $2[Si_3O_9]-O$.

An examination of Fig. 17c, d reveals that although there are only four Si tetrahedra per link of the amphibole ribbon, the characteristic feature of this ribbon is the six-membered ring, and in exactly the same way, six Si tetrahedra per link of a xonotlite ribbon produce an eight-membered ring (the pairs of Si bridge tetrahedra are counted only once in each link).

The structure of xonotlite (Fig. 18a) differs greatly, however, from those of pyroxene and amphibole. It contains only two stories, one made up of a continuous "wall" (and not separate ribbons) of Ca polyhedra, and the other of $[Si_6O_{17}]_\infty$ oxygen-silicon ribbons, identical to each other (and not in two orientations, as in amphiboles, and even in wollastonite).

Just as in amphiboles, where the O atoms lost from the $[Si_4O_{11}]_\infty$ chains on condensation are hydrated and resulting OH groups are attached to the silicate chain, as in tremolite, $Ca_2Mg_5[Si_8O_{22}](OH)_2$, so we get hydroxyl groups added to the $[Si_6O_{17}]$ radical in xonotlite, $Ca_6[Si_6O_{17}](OH)_2$.

As we have stated, the great majority of hydrated Ca silicates (cement silicates) are transformed at high temperatures to orientated fibers of wollastonite. Almost all these hydrated silicates are either themselves fibrous, or are platy but easily split into fibers, and along the axis of the fiber they have a wollastonite (xonotlite) repeat distance of ~ 7.2-7.3 Å, i.e., equal to twice the length of an edge of a Ca octahedron.

At low temperatures, a proportion, at least, of these compounds contain xonotlite ribbons, and for this reason this ribbon can be related to the appearance of cementitious properties in the compounds. As well as in xonotlite, the structure of which (Fig. 18a) has been fully analyzed [9], xonotlite ribbons are supposed in foshagite, hillebrandite, and different tobermorites. In some cement silicates the xonotlite ribbons are present as independent radicals, while in others they are condensed, although still retaining their individuality (Vernadskii's "radicals").

The structures of xonotlite, foshagite [27], and hillebrandite [17] have so many features in common, that we have placed them together on Fig. 18 (a, b, c). If the cell dimensions of the xonotlite structure are a, b, and c, then foshagite has the same b and c, and hillebrandite has the same a and b. Both the latter structures are produced by the process of liming, i.e., adding lime to xonotlite, introducing portlandite, $Ca(OH)_2$ into the structure. In foshagite the additional amount of $Ca(OH)_2$ added is small

$$Ca_6[Si_6O_{17}](OH)_2 + 2Ca(OH)_2 = Ca_8[Si_6O_{17}](OH)_6$$
$$\text{xonotlite} \qquad \text{portlandite} \qquad \text{foshagite}$$

As shown in Fig. 18b, the "molecules" of $Ca(OH)_2$ are introduced in the form of extra Ca octahedra between the xonotlite ribbons, increasing the a dimension but leaving the other two unchanged.

The effect of heating foshagite is completely obvious. From the equation

$$Ca_8[Si_6O_{17}](OH)_6 \rightarrow 6CaSiO_3 + 2CaO + 3H_2O\uparrow$$

we see that it breaks down into oriented wollastonite fibers (parallel to the b-axis of foshagite, and free lime.

The facts are somewhat different for hillebrandite, the most simply prepared hydrated cement silicate. The a and b dimensions are the same as in xonotlite, but c is the sum of the c dimensions of xonotlite and portlandite, $Ca(OH)_2$. In view of this sum we can assume that the chemical formula of hillebrandite is given by the equation

$$Ca_6[Si_6O_{17}](OH)_2 + 6Ca(OH)_2 = 6\{2CaO\cdot SiO_2\cdot H_2O\} + H_2O\uparrow$$
$$\text{xonotlite} \qquad \text{portlandite} \qquad \text{hillebrandite}$$

This simple equation should not be taken to imply that there actually exist two sorts of molecules in hillebrandite. Figure 18c shows a somewhat simplified model of hillebrandite, and a comparison of this with xonotlite (Fig. 18a) will explain what we mean. In xonotlite, the continuous layer or wall of Ca polyhedra is shared simultaneously by two layers of the xonotlite oxygen-silicon ribbons. In hillebrandite, the xonotlite ribbon layer has its own unshared layers of Ca polyhedra both above and below it. The corners of the Ca polyhedra pointing toward the xonotlite ribbon are O atoms, while those on the other side are OH groups, and it is these OH groups, like those in $Ca(OH)_2$ and $Mg(OH)_2$ [28], which hold together the xonotlite packets by hydrogen bonding. This weak bonding is the reason for the perfect cleavage of hillebrandite. Figure 18c can be adapted to the following form:

$$
\left.
\begin{array}{cccc}
\text{Ca} & \text{Ca} & \text{Ca} & \text{Ca} \\
\text{Si}_6\text{O}_{17} & \text{Si}_6\text{O}_{17} & \text{Si}_6\text{O}_{17} & \\
\text{Ca} & \text{Ca} & \text{Ca} & \text{Ca}
\end{array}
\right\} \text{xonotlite}
$$

$$
\left.
\begin{array}{cccc}
\text{OH} & \text{OH} & \text{OH} & \\
\hline
\text{OH} & \text{OH} & \text{OH} &
\end{array}
\right\} \text{portlandite}
$$

$$
\left.
\begin{array}{cccc}
\text{Ca} & \text{Ca} & \text{Ca} & \text{Ca} \\
\text{Si}_6\text{O}_{17} & \text{Si}_6\text{O}_{17} & \text{Si}_6\text{O}_{17} & \\
\text{Ca} & \text{Ca} & \text{Ca} & \text{Ca}
\end{array}
\right\} \text{xonotlite}
$$

Thus the "molecules" of portlandite shown in the formula of hillebrandite turn out, in actual fact, to be two "half-molecules," $Ca_{0.5}OHHOCa_{0.5}$, between which the cleavage of portlandite passes.

If the intermediate portlandite layer contains $2\cdot 0.5 + n$ molecules of $Ca(OH)_2$, then we can talk about intergrowth of hillebrandite with $Ca(OH)_2$, quite a frequent occurrence.

Hillebrandite is one of the few hydrated Ca silicates which, on heating, does not give wollastonite, but unoriented crystals of a calcium orthosilicate. This process takes place in two stages. Just as portlandite begins to dissociate at 550°, so does hillebrandite begin to break down at the same temperature, according to the equation

$$Ca_6[Si_6O_{17}](OH_2)\cdot 6Ca(OH)_2 \xrightarrow{550°} Ca_6[Si_6O_{17}](OH)_2 + 6CaO + 6H_2O\uparrow.$$

At higher temperatures, instead of the oriented xonotlite fibers, we are left with disordered grains of Ca orthosilicate

$$Ca_6[Si_6O_{17}](OH)_2 + 6CaO \rightarrow 6Ca_2[SiO_4] + H_2O\uparrow.$$

A large number of the thermal and X-ray characteristics of many hydrated Ca silicates have been collected and published, under the editorship of J. Bernal, in a special small handbook on Ca hydrosilicates [29], which is of exceptional value to all investigators in the cement field.

The Crystal Structure of Rhodonite — $MnSiO_3$

A large number of mineral handbooks avoid the question of the structure of rhodonite, although a few ascribe to this mineral $[Si_3O_9]$ rings of the "benitoite" type, analogous to those very probably present in pseudohexagonal α-wollastonite, and also in $BaSiO_3$ and $RbBeF_3$ [30]. It appears that in the case of $MnSiO_3$, a three-membered ring is characteristic of the high-temperature α-modification, which unlike α-wollastonite, however, is thermodynamically unstable; it will nonetheless be formed from a melt of $MnSiO_3$, in accordance with Ostwald's rule of phases, and under favorable conditions the finest crystals will be preserved even down to ordinary temperatures [31].

The low-temperature modification is β-$MnSiO_3$, natural rhodonite, and according to results shown in Table 3 it has a structure of its own, different from that of β-$CaSiO_3$, although it will accept up to 20% Ca in exchange for Mn. Bustamite, $CaMn[SiO_3]_2$, with approximately equal quantities of Ca and Mn, crystallizes in the β-wollastonite structure.

In 1954 F. Liebau [32] showed that rhodonite, which had a unit cell containing an unusual number, 10, of $MnSiO_3$ molecules, possessed yet another new type of $[SiO_3]_\infty$ metasilicate chain, a chain with five-membered links, shown in Fig. 14c. Subsequently the same author showed [33] that pyroxmangite, $(Mn, Fe, Mg, Ca)SiO_3$, could be expected to have $[SiO_3]_\infty$ chains with even longer links, seven-membered ones (Fig. 14d). A comparison of these chains with pyroxene and pyroxenoid chains (Fig. 15a,b) shows that the rhodonite and pyroxmangite chains are made up of pieces of the first two chains.

The cause of the unexpected shape of the rhodonite type metasilicate chain was explained by Kh. S. Mamedov [16] from the application of some ideas on the close relationship between the geometry of oxygen-silicon radicals and the singular properties of the "principal" cations which form the skeleton of the structure.

The $Mn^{..}$ cation has a d electron shell which is exactly half full (5 electrons out of 10), and it possesses a number of special features; in particular, the possibility of increasing its valence, which can bring about a nearest-neighbor interatomic distance of much less than the 0.91 Å accepted as the ionic radius.

In connection with this, it is found that the columns of Mn octahedra, the analogs of the Ca columns, differ from the latter in that they are all (except one octahedron) formed into nine-membered links, which are joined at the side to each other in such a way, that along the long (b) axis of the crystal, there is formed an infinite, "rasped" ribbon. The middle three octahedra of each link are connected with the leading three of the link behind and the trailing three of the link in front (Fig. 19).

TABLE 3. Unit cell constants in the $CaSiO_3$ — $MnSiO_3$ system

Mineral and formula	a, Å	b, Å	c, Å	α	β	γ
Wollastonite - $CaSiO_3$	7.88	7.27	7.07	90°00'	95°16'	103°22'
Bustamite - $CaMn[SiO_3]_2$	7.66	7.17	6.88	92°08'	94°54'	101°35'
Rhodonite - $MnSiO_3$	7.77	12.20	6.70	85°15'	94°00'	111°29'
Pyroxmangite - $(Mn,Fe)SiO_3$	7.50	17.20	6.82	82°48'	94°20'	113°17'

The repeat distance (b = 12.20 Å) of this corrugated ribbon can be seen from Fig. 19. The center of symmetry of the unit cell lies in the fifth octahedron of each nine, and there is another center in the omitted (empty) octahedron (there are two further symmetry centers, which will be left to the reader to discover). In the arrangement shown, a unit cell with origin in the omitted octahedron has been outlined, and it can be seen that this cell contains only 9 octahedra, although we stated above that the complete unit cell contains 10 "molecules" of $MnSiO_3$.

In the structure of rhodonite, as in other pyroxenes and pyroxenoids, a story of cation octahedra alternates with a story containing Si tetrahedra. In rhodonite, however, this middle tetrahedron layer also contains the tenth Mn octahedron, which joins together fifth (middle) octahedra in the octahedron layers, in the form of a letter H lying on its side (Fig. 20). These Mn atoms in the tenth octahedra also lie at centers of symmetry.*

To the rasped ribbon (which like other chain metasilicates has a "mean" thickness of 3 octahedra) is attached a metasilicate oxygen-silicon chain, with the normal formula of $[SiO_3]_\infty$, oriented in the same general direction as the basic ribbon of Mn octahedra.

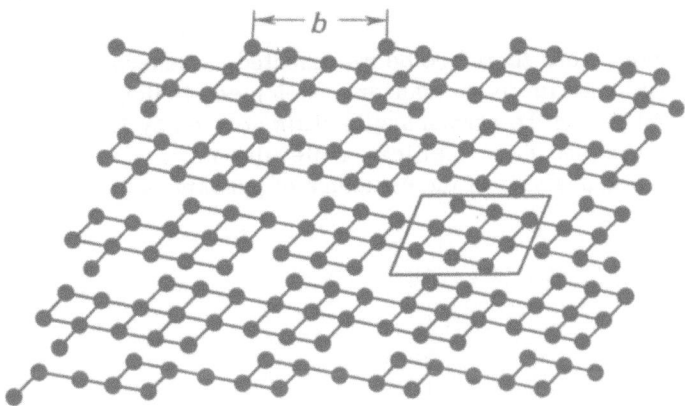

Fig. 19. Corrugated (rasped) ribbons of Mn atoms in the structure of rhodonite. The central triplet of each set of nine Mn atoms is held between the leading triplet of the set of nine behind and the trailing triplet of the set in front. The b × c unit cell of rhodonite is outlined.

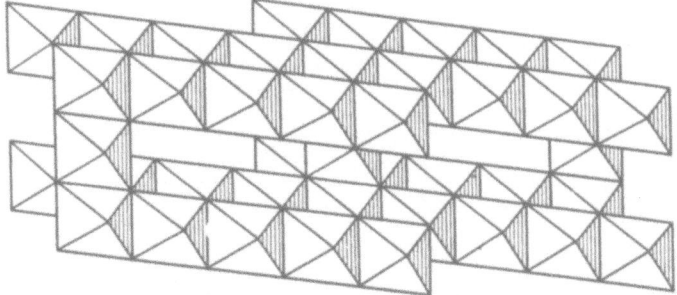

Fig. 20. Coupling of sets of nine Mn octahedra in the odd stories of the rhodonite structure, by single Mn octahedra in the even stories.

*It has been stated above that the Ca content in rhodonite can reach 20%; normally it is about 10% [34] which corresponds to the special position of the tenth octahedron in the structure. The other 10% will go into the other centrosymmetric octahedron, although, of course, this is less favorable.

Figure 16c shows how the $[SiO_3]_\infty$ chain passes along the second, third, and fourth Mn octahedra (counting from the left), according to the normal wollastonite rule (see Fig. 16b); a pair of tetrahedra (Si_2O_7 group) is attached to the second octahedron, a single tetrahedron to the third, and again a pair to the fourth; there then occurs (Fig. 16c) a jump across to the second octahedron of the neighboring (nine-membered) link of Mn octahedra. When this jump occurs, the Si_2O_7 group on the fourth octahedron of the first Mn link is attached directly to a further Si_2O_7 group, on the second octahedron of the neighboring Mn link, which means that a SiO_4 group has been left out; with the second Mn octahedron of the second link, construction is begun of the next five-membered link of the oxygen-silicon chain.

In the fifth octahedron of each Mn link, there is, as we have stated, a center of symmetry. So if the oxygen-silicon chain we have traced out passes over the second, third, and fourth octahedra, with the Si tetrahedra pointing downward, then there must be another five-membered oxygen-silicon chain passing under the sixth, seventh, and eighth octahedra of the same set of nine, with Si tetrahedra pointing upward.

The formula of the rhodonite oxygen-silicon chain can be written as $[Si_5O_{15}]_\infty$, or more precisely, as $[Si_{2+1+2}O_{15}]_\infty$.

The structure of rhodonite can therefore be represented as separate pieces of a wollastonite chain, each three octahedra long with five tetrahedra; each piece is displaced by one step from the next. The seven-membered link in the pyroxmangite chain derived by Liebau (from the unit cell dimensions and the number of Mn and Fe metasilicate molecules in it) can be written in the form $[Si_{2+2+1+2}O_{21}]$. A more detailed analysis of the structure of pyroxmangite has not yet been produced[33].

Oxygen–Silicon Networks Based on Wollastonite
Chains and Xonotlite Ribbons

According to the number of tetrahedra contained in each link (repeat distance) of a $[SiO_3]_\infty$ chain, German authors use a nomenclature under which a pyroxene chain is called a Zweierkette, a wollastonite chain a Dreierkette, a rhodonite chain a Fünferkette, and a pyroxmangite chain a Siebenerkette.[*] The last two names are not particularly essential, but if we use the first two, we can then speak of amphibole and xonotlite ribbons as Zweierdoppelkette and Dreierdoppelkette,[†] and the corresponding two types of network as Zweierschicht and Dreierschicht.

The first type of network we will deal with is the familiar one found in the structures of talc, pyrophyllite, kaolin, micas, and chlorites. It arises by side-condensation (sharing of outer O atoms) of amphibole ribbons, lying in line and related by planes of symmetry between them (talcization of amphiboles); the condensation is continued to infinity along the direction perpendicular to the axis of the ribbons. In the process, "extra" amphibole ribbons are formed (Fig. 21), which are crammed between adjacent original ribbons. After condensation it is not possible to distinguish between the "extra" and the original ribbons. The whole network is made up entirely of hexagonal rings or loops. The simplicity of the network produced is due to the geometrical properties of regular hexagons, which can completely fill an area, leaving no gaps. This is not possible with octagons, and so the corresponding

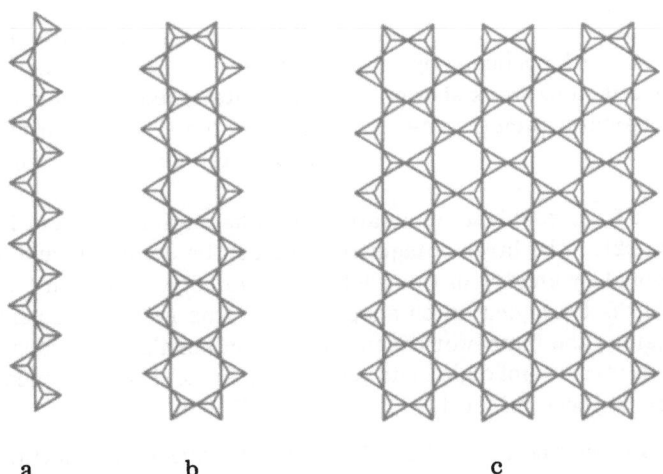

a   b   c

Fig. 21. Flat oxygen–silicon networks with six-membered rings (c) formed as a condensation product of pyroxene chains (a) via amphibole ribbons (b).

---

[*] The name "Viererkette" is given to the batisite chain $[Si_4O_{12}]$ (see footnote on p. 15).
[†] In Fig. 17, in addition to Zweier(Dreier)ketten and Zweier(Dreier)doppelketten, we also show an Einerkette and an Einerdoppelkette (Fig. 17a,b). The latter is found in the structures of sillimanite, Al_2SiO_5 [18] and mullite.

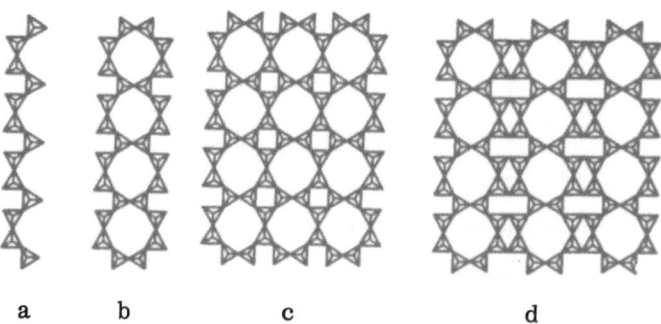

a   b   c   d

Fig. 22. Two condensation products of pyroxenoid chains (a) via xonotlite ribbons (b): the idealized apophyllite network, with extra Si_2O_7 bridge groups (c), and a theoretical arrangement without bridge groups (d).

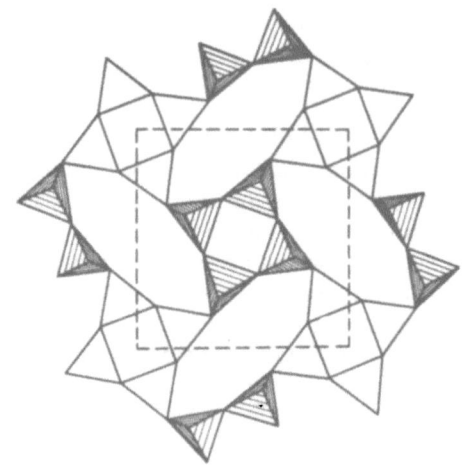

Fig. 23. Apophyllite network produced by compression of the idealized network shown in Fig. 22c, with tetragonal symmetry retained.

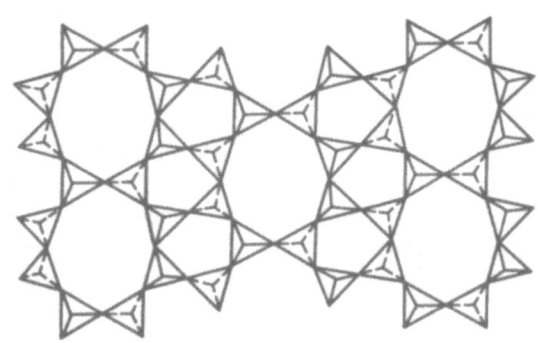

Fig. 24. Okenite network of oxygen-silicon tetrahedra, arising when the condensation of xonotlite ribbons is carried on by a glide plane (eight-membered rings alternate with double the number of five-membered rings).

xonotlite ribbon network—a Dreierschicht—normally exists as an arrangement in which octagons alternate with rectangles (Fig. 22). The large octagonal rings can be deformed and compressed, but in such a way that the tetragonal symmetry of the whole is retained. As a result we derive the network (Fig. 23) which is found in the tetragonal mica apophyllite. This network is rare, but if it is condensed in the third dimension, the framework found in such minerals as feldspars, scapolites, and zeolites is produced, and alternation of eight- and four-membered rings is a particularly characteristic feature of the structure of these minerals.

Apophyllite networks cannot be created by simple condensation of xonotlite ribbons, because eight-membered rings do not alternate with four-membered ones in the latter.*

If, however, xonotlite ribbons are condensed with simultaneous gliding, a network is derived in which eight-membered rings alternate with twice the number of five-membered ones, without bridge groups being introduced (Fig. 24). This network is found (on moving outside the limits of the unit cell)

---

*But it is easily made when we condense the batisite chains $[Si_4O_{12}]\infty$.

in layered Ca silicates such as okenite, truscottite, gyrolite, and others [35].* In all the Zweier-schichten and Dreierschichten we have shown, the general formula is the same, $[Si_2O_5]_{\infty, \infty}$.

In two of the Dreierschicht arrangements (Figs. 22d, 24) there is one eight-membered ring for every two other rings, not eight-membered. It appears that, in accordance with what we said above about xonotlite ribbons, eight-membered rings are favorable for cement structures with large cations; and in the common mica-like tobermorites [36], a network is constructed out of these same xonotlite ribbons, also a Dreierschicht, in which for one four-membered ring plus one six-membered ring there are three eight-membered rings. This is achieved by arranging and condensing the xonotlite ribbons on two levels, which results in a corrugated or pleated network. Details of the pleating can be seen from Fig. 25a, b, which shows a basal section (cutting the xonotlite ribbons) of three layers, and a view of the network from above. Unfortunately, these views do not show as well as the three-dimensional model does how "extra" ribbons are created lying perpendicular to the plane (Fig. 25b), one consisting of alternating six-membered and four-membered rings, the other of eight-membered rings only. The relationship given above between the numbers of eight-membered and noneight-membered rings is derived from this.

When one ring is displaced over the other, the lattice constant in the plane of the xonotlite ribbon and perpendicular to its axis is decreased to two thirds its original value. The two halves of each eight-membered ring participate in the condensation process in different ways; one half uses both its exterior tetrahedra, and the other half uses only its bridge tetrahedron (Fig. 25b). Of the six tetrahedra per ring, only three have their corners shared; this results in a somewhat unusual formula for the tobermorite network

$$2[Si_6O_{17}]_\infty - 3O = [Si_{12}O_{31}]_{\infty, \infty}.$$

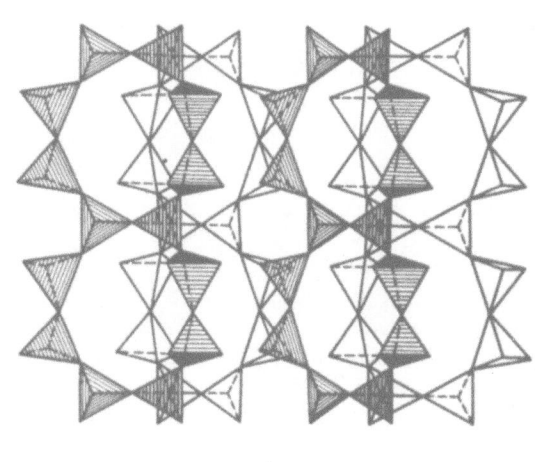

Fig. 25. Structure of tobermorite: (a) basal projection (looking along the axis of xonotlite ribbons, condensed into a pleated network); (b) the method of pleating — mutual partial superposition of the xonotlite ribbons (circles represent Ca cations, the molecules of interlayer water are not shown).

_____

*Recently we have shown that this kind of network really exists in the structure of eudidymite (E. A. Pobedimskaya, N. V. Belov, Doklady AN SSSR, 136, No. 6, 1961).

Having established this oxygen-silicon radical, we are able to give the experimental formula of tobermorite a rational appearance (after doubling)

$$2\{5CaO \cdot 6SiO_2 \cdot xH_2O\} = Ca_{10}[Si_{12}O_{31}](OH)_6.$$

We have written the constant part of the formula of tobermorite. Tobermorites characteristically contain large quantities of water, and the number of water molecules increases in stages, in step with a simultaneous increase in the c lattice parameter, a and b remaining unchanged, as follows:

$$9 \text{ Å-hydrate } Ca_{10}[Si_{12}O_{31}](OH)_6 \cdot 3H_2O \qquad c = 18 \text{ Å}$$
$$11 \text{ Å-hydrate } Ca_{10}[Si_{12}O_{31}](OH)_6 \cdot 8H_2O \qquad c = 22 \text{ Å}$$
$$14 \text{ Å-hydrate } Ca_{10}[Si_{12}O_{31}](OH)_6 \cdot 18H_2O \qquad c = 28 \text{ Å}$$

The number of angstroms quoted before the word "hydrate" is half the c lattice parameter, and corresponds to the interplanar spacing d_{002} of the corresponding line on the X-ray powder photograph by which the three basic tobermorites are always distinguished. The other lines are mostly governed by the a and b parameters, and their positions on the powder photograph remain practically unchanged. This provides evidence that both the Ca atoms (shown by circles) and the OH groups (not shown) are more or less fixed in position around the pleated silicon-oxygen network.

At high temperatures ($\sim 750°$) tobermorites lose water and change into oriented fibers of wollastonite, as do xonotlite and foshagite. The "xonotlitic nature" of tobermorite is demonstrated by the similarity of the infrared absorption spectra of the two minerals in the 8-15 micron region. Tobermorite, however, has a characteristic band at 6.2 microns which is normally related to the interlayer water, and which in fact is absent on the xonotlite spectrum.

The platy hydrated Ca silicates which we have examined, in accord with the way they were built up geometrically from xonotlite ribbons, give on heating oriented fibers of wollastonite, which would be impossible for an apophyllite network.*

---

*A most interesting structure built by xonotlite ribbons is that of the very active molecular sieve - mordenite (W. M. Meier, Z. Krist. 115, 439, 1962).

Molecular Sieves

We will now pass on from the second "chapter" flat networks, the Dreierschichten, characterized by eight-membered rings and oxygen-silicon radicals infinite in two directions, to their three-dimensional analogs—frameworks made up of Si_2O_7 groups (mostly, of course, with additional "connecting" SiO_4 groups). Frameworks of this type (Fersman's "braids") are found in the structures of feldspars, scapolites, and zeolites, although in these cases, a considerable proportion of the Si tetrahedra have been replaced by somewhat larger Al tetrahedra. From one point of view, these are the oldest representatives of the second "chapter," dating from the time when separation of silicate crystal chemistry into two "chapters" had not yet been spoken of. On the other hand, a great deal of new work on these structures has been carried out in recent years. We will dwell here only on the zeolites, which under the name of "molecular sieves" now hold the center of attention in applied mineralogy, chemical technology, and preparative chemistry [37].

The simpler zeolites have been used for a long time in water softening; these include hydrosodalite (ultramarine), cancrinite, and analcite, which in the dry, dehydrated state are excellent absorbents in the separation, from mixtures, of the simpler molecules, such as H_2O, NH_3, and others of diameter ~ 3 Å. This selective absorption property is governed by the diameter of the apertures in the beryl-type six-membered oxygen-(aluminum-)silicon rings from which these zeolites are constructed.

Figure 26 is a detailed diagram of the framework structure of hydrosodalite, consisting of open cages of Si, Al tetrahedra (I like to call them "Chinese lanterns") containing alkali or alkaline earth cations and water molecules (not shown). Leading into each lantern are six smaller apertures in the cube face positions, consisting of four-membered rings, and eight larger apertures in the octahedron face positions, consisting of the six-membered rings we mentioned above.

Figure 26 shows how the "lanterns" are joined together to form the cubic structure of hydrosodalite, and it can be seen that the floor of one lantern serves as the roof of the lantern below it, and that all six four-membered rings in each lantern are shared simultaneously with neighboring lanterns. The same thing is true of the six-membered rings.

The hydrosodalite "lantern" has turned out to be the fundamental constructional feature—the ready-made building unit—of a large proportion of modern molecular sieves, i.e., natural and synthetic zeolites. In many of the articles devoted to this subject, the lantern has been shown as a simple geometrical figure, the cubo-octahedron, with six small square faces and eight larger hexagonal faces. The corners of this cubo-octahedron correspond to the centers of the tetrahedra, i.e., to the atoms of Si(Al). This cubo-octahedron is exactly the same as the well-known Fedorov parallelohedron (heptaparallelohedron), dealt with in all Soviet crystallography courses; such parallelohedra can completely fill a space, leaving no gaps. If eight of these cubo-octahedra are placed at the corners of a cubic cell in such a way that they share square faces, a hollow is formed inside the cell of exactly the same shape as the original cubo-octahedron so that it can be filled with a ninth cubo-octahedron (Fig. 27). The latter is bounded by the hexagonal faces of the original cubo-octahedrons since each six-membered ring is shared between two hydrosodalite lanterns.

As we have said, the maximum diameter of the molecules which can be absorbed by hydrosodalite and similar sieves is ~ 3 Å. This figure may be slightly adjusted in one direction or the other by

Fig. 26. The oxygen-silicon framework in the structure of hydrosodalite (ultra-marine), consisting of open "Chinese lantern" cages made up of a central hollow surrounded by 24 tetrahedra distributed as six- and eight-membered rings (a,b). At the center of the cube formed by eight such "lanterns" (the four front ones only are shown in Fig. 26c) is formed another geometrically identical "lantern," the walls of which are shared with the eight corner "lanterns."

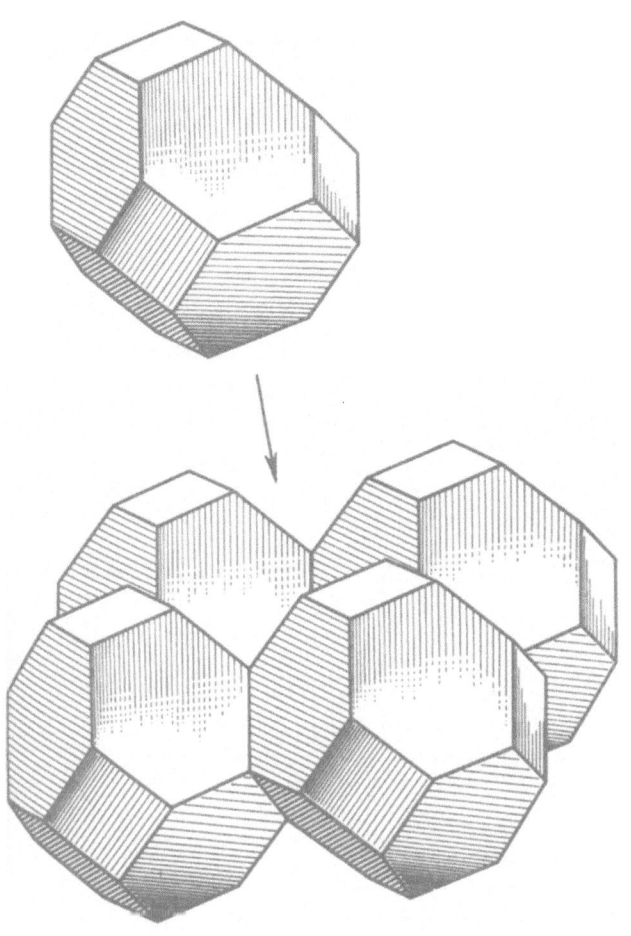

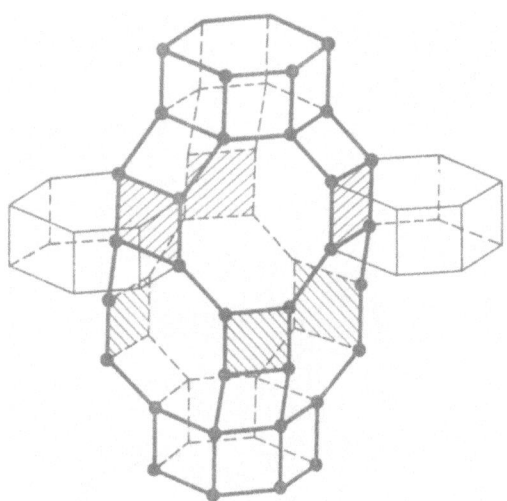

Fig. 28. The oxygen-silicon framework in the rhombohedral structure of chabazite. Only the Si(Al) atoms at the centers of the tetrahedra forming the two-storied milarite rings are shown. These appear here as hexagonal prisms. The shaded squares indicate the places where the hexagonal prisms butt up against the vast central hollow, through their square side faces. Sets of four six-membered rings form a large eight-membered window, common to two hollows, and permitting passage and absorption of large molecules up to about 5 Å in diameter.

Fig. 27. Heptaparallelohedra (Fedorov's cubo-octahedron with 14 surfaces) filling a space, leaving no gaps, in a body-centered cubic lattice.

the polarity and other properties of the absorbed molecules which may "activate" the molecule, helping it to pass through the narrow "window." It has been known for a long time that molecules of argon, with a cross section of 3.84 Å, and even molecules of methane (4.25 Å), would without difficulty pass through and be absorbed in the lattice of chabazite. Propane (4.9 Å) and n-butane are absorbed with difficulty, and isobutane cannot be absorbed. This high upper limit was for a long time a great puzzle, and people tried to devise a rhombohedral structure for chabazite which was made up of hexagonal (beryl) rings, related according to rhombohedral symmetry and lying in parallel orientations. It was not difficult to prove from models that a structure with larger apertures could not be produced in this fashion. The riddle was solved in 1958, when it was shown [38] that chabazite was made up, not of single-storied $[Si_6O_{18}]$ rings of the beryl (dioptase) type, but of the $[Si_{12}O_{30}]$ milarite rings, also six-membered but two-storied, which we discovered in 1949 [4], and which we classified above as second-"chapter" radicals. If we pack these rings together remembering the rhombohedral symmetry, we arrive at a structure (Figs. 28, 29) which has hollows bounded by four- and eight-membered rings as well as the two six-membered rings (in Fig. 28 the two-storied rings are shown by hexagonal prisms, the twelve corners of which correspond to the Si(Al) atoms lying at the centers of the oxygen tetrahedra.) The eight-membered rings have a diameter of ~5 Å and make chabazite an effective sieve for the larger molecules.

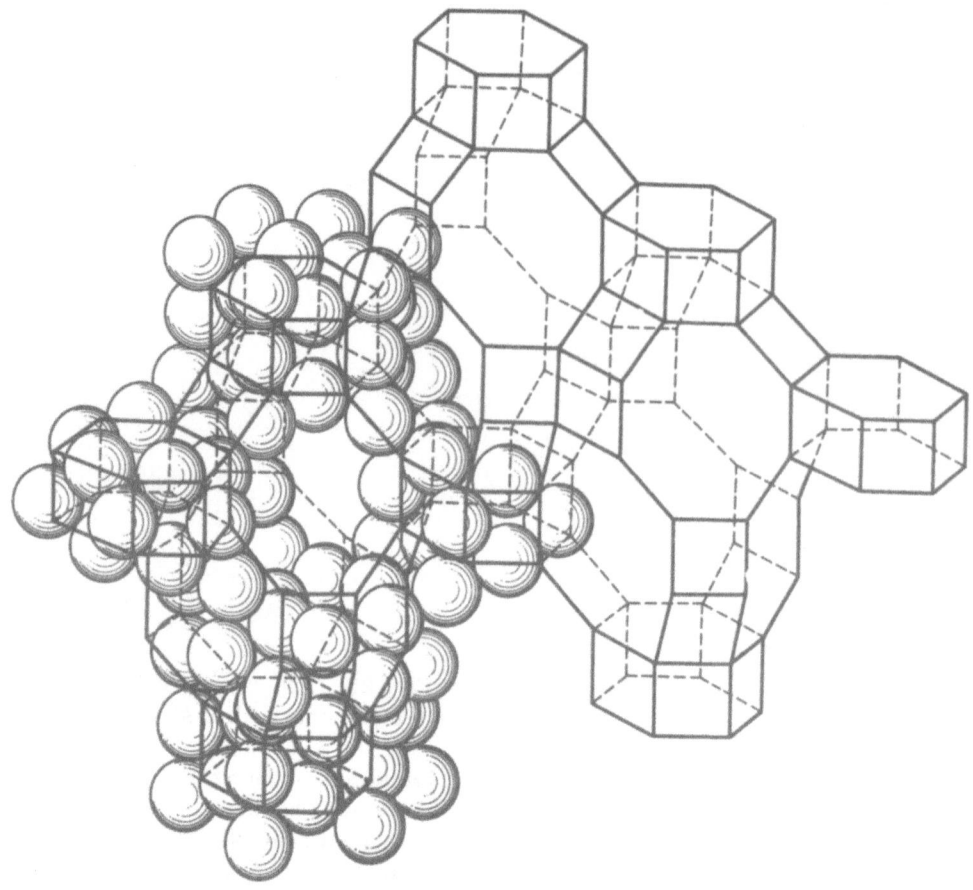

Fig. 29. Three interconnecting hollows in the chabazite structure. In one of them the oxygen atoms are shown in full to give a true representation of the diameter of the apertures, the eight-membered rings.

Attempts to synthesize chabazite have been unsuccessful up to the present. It was therefore an important event in the history of molecular sieves when a very easily synthesized sieve, similar to hydrosodalite, was discovered, which instead of ordinary four-membered rings, had two-storied four-membered rings, and was just as effective a sieve as chabazite [39, 40].

Single-storied four-membered rings are present in the simple hydrosodalite model as the square faces shared between two cubo-octahedra. If the four-membered rings are made two-storied, they appear in the model as square prisms, held between pairs of cubo-octahedra. Figure 30 shows very well how a large hollow is formed inside eight cubo-octahedra joined in this fashion, a hollow of much larger volume than that inside a cube of hydrosodalite "lanterns." The lanterns in the synthetic zeolite, which is known as Linde sieve A [41], deviate from this model slightly, mostly as regards the small hollows inside the square prisms.

The largest windows in the central hollow of Linde sieve A (Fig. 30) are also eight-membered, and as in chabazite, there are six of them. Both structures have the same number of four-membered windows. Chabazite has two six-membered windows, while sieve A has six.* Thus the upper size limit for absorbed molecules is the same in sieve A as in chabazite (5A), but the volume of the hollow is much greater in sieve A. This volume is usually "measured" in terms of the number of water molecules which can be placed in the hollow. For chabazite it is ~ 12 H_2O, while for sieve A it is ~ 24 H_2O.

*In Fig. 30 the cubo-octahedra are not shown in the conventional solid form, but as frameworks, in order to show how the original six- and four-membered rings are distributed over the faces.

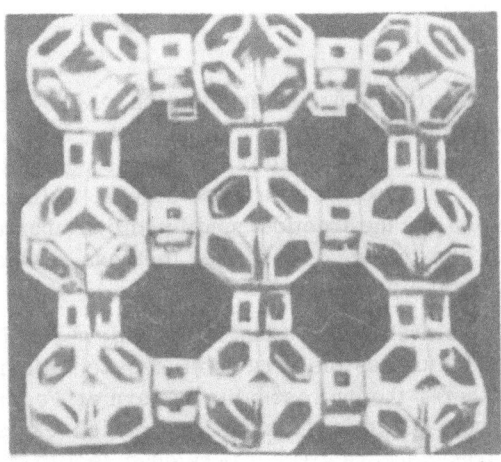

Fig. 30. Linde sieve A — a primitive cubic structure of Fedorov cubo-octahedra (hydrosodalite lanterns) and tetragonal prisms with a large central hollow, entrance to which is gained principally through six large windows (eight-membered rings).

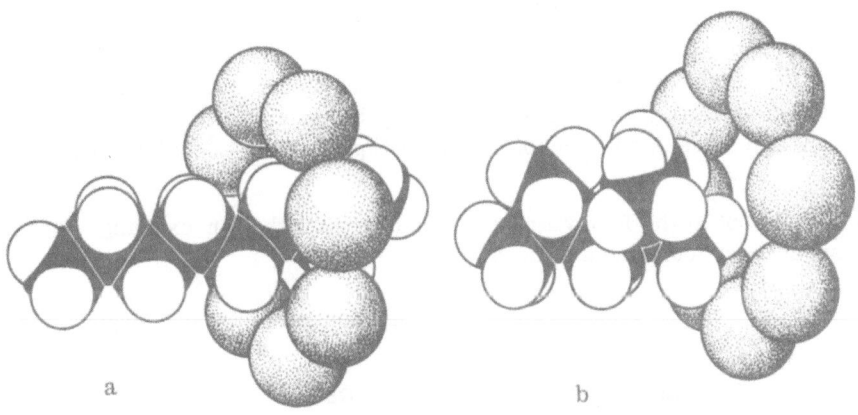

a   b

Fig. 31. Separation of gasoline octanes using eight-membered silicon-oxygen rings: (a) n-octane molecule; (b) iso-octane molecule.

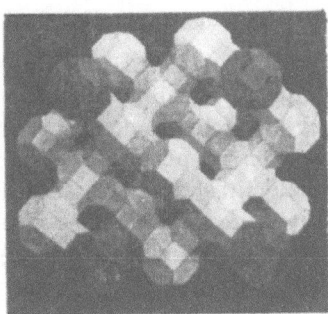

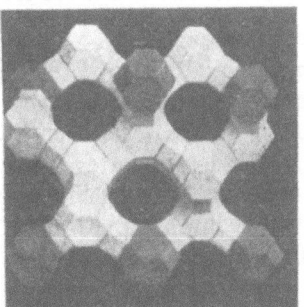

Fig. 32. The cubic structure of Linde sieve X, consisting of Fedorov cubo-octahedra connected by hexagonal prisms. Each cubo-octahedron is surrounded by only four others, as in the diamond structure.

Fig. 33. The wide twelve-membered windows in sieve X. The windows are inclined to [110], the channel axis, which therefore has an elliptical cross section.

Figure 29 shows the eight-membered window leading into the hollow in chabazite in its "natural" form: it is bounded by eight spheres of oxygen atoms. Figure 31a, b shows how such a window can separate thin n-octane molecules from shorter, fatter iso-octane molecules. This is an example of one of the most effective operations carried out with these sieves, in the separation from gasoline fractions of the very high quality iso-octane fraction, undiluted by impurities which would lower the quality. This operation could not be performed by earlier methods, at least not economically, because both octanes have very similar physical properties.

Even more effective is Linde sieve X, which has large hollows, and more important, particularly large apertures: this is found in nature as the rare cubic zeolite faujasite [42]. This possesses the same constructional unit as the others, the hydrosodalite lantern or Fedorov cubo-octahedron, connected via two-storied milarite rings shown in the simplified models as hexagonal prisms. The reason why this method of connecting cubo-octahedra gives a very open zeolite framework, with very large hollows and wide windows, will be clear if we examine carefully the structure of the hydrosodalite "lantern" (Fig. 26). Its envelope is made up of 24 (Si, Al) tetrahedra, lying on six separate four-membered rings or on eight six-membered rings.

As regards the six-membered rings, each tetrahedron is simultaneously part of two of these. From Fig. 26 it can be seen that in half the six-membered rings the component tetrahedra have free oxygen corners pointing outward. It will be apparent that only faces of the first type can be coupled in pairs via the oxygen atoms, and so in coupling by means of hexagonal prisms each cubo-octahedron will have only four others around it, and not eight.

The framework of the cubic zeolite faujasite (synthetic sieve X), will therefore be constructed on the principle of the diamond lattice, giving [43] a very open structure with extremely big hollows (Figs. 32, 33), the apertures of which are twelve-membered rings, capable of passing a large-diameter molecule. While sieve A will accept molecules up to 4 and 5 Å, sieve X will take them up to 10 and 13 Å.

This is brought out in the following table, which shows the ion-exchange capabilities of these sieves with respect to various-sized cations.

	Exchanged	Not exchanged
Ultramarine (sodalite)	K	Cs
Analcite, basic cancrinite	Rb	Cs
Chabazite	Cs	$N(CH_3)_4$
Faujasite (sieve X)	$N(CH_3)_4$	$N(C_2H_5)_4$

This table has been borrowed from the work of R. M. Barrer [40], who was a pioneer in the introduction of zeolites into chemical technology, and who is still the foremost specialist in the subject.

Modern molecular sieves, synthesized at controlled temperatures and pressures, are stable up to $\sim 800°$. They lose all their water a good deal before this, becoming particularly effective absorbents of traces of moisture, which may be extremely dangerous, for example, in compressed gases. The absorption properties of sieves are, of course, partly dependent on the chemical properties of the absorbed molecules: the sieves are most effective with polar molecules, and also unsaturated hydrocarbons. For any particular molecule there is a complex but definite relationship between absorption and thermodynamic factors, and if the appropriate functions are known a great many separations become feasible [44].

One very important point is that it is possible to alter the absorption properties of a sieve by preliminary ion exchange. Since up to half the Si tetrahedra in the framework may have been replaced by Al tetrahedra, cations must be present in the sieve. Thus the formula of ordinary sodalite is $3Na_2Al_2Si_2O_8 \cdot 2NaCl = Na_8[Al_6Si_6O_{24}]Cl_2$, and that of hydrosodalite $3Na_2H_2Si_2O_8 \cdot 2NaOH \cdot nH_2O$. Cation exchange does not take place atom for atom, but according to classical valence principles

$$2Na^{1+} \rightarrow Ca^{2+},$$

i.e., eight Na atoms in sodalite are replaced by only four Ca atoms in Ca-sodalite. These cations position themselves in the windows of the hollows, and if their number is reduced it greatly increases the permeability of the sieve. The size limits of 4 and 5 Å quoted above correspond to Na-sieve A and Ca-sieve A, respectively.

Surprisingly enough, in X sieves the position is reversed; the limits for Na-sieve X and Ca-sieve X are 13 Å and 10 Å, respectively. It appears that on decreasing the number of cations, in this case, they are drawn from the double six-membered windows, which have little effect on the absorptive ability of the sieve. At the open windows, replacement of univalent Na by bivalent Ca sharply increases the retarding effect of cationic shielding.

We have already said how easy it is to synthesize many zeolites, in particular the A and X sieves, which are precipitated on reacting $NaAlO_2$ solutions with sodium silicate at low temperatures and pressures (and controlled pH). Chabazite has not been synthesized up to the present, however. An interesting explanation of this can be given in terms of the second "chapter" principles.

We have emphasized above that the Si_2O_7 group is mostly found in the shape of a trigonal prism, rather than as a centrosymmetric octahedron (twisted prism). The Si^{4+} ion in its tetrahedron is so inert that it submits to the tendency of the O atom to have two valence bonds at the tetrahedral angle of 109° to each other. (It is usually accepted that the bonds in the SiO_4 tetrahedron are at least half covalent.) This tendency is opposed by the obvious inclination of the highly-charged Si^{4+} ions to repel each other, which would lead to a Si—O—Si angle close to 180°.

The compromise solution is an angle between these two, of about 140°. To inscribe a Si_2O_7 group with this angle in the octahedron is quite difficult, and impossible without losing the center of symmetry, but the desired configuration can be achieved quite easily if the Si_2O_7 group is inscribed in a trigonal prism, in which the bridge O atom (slightly attracting both Si atoms toward it) is slid along the central plane, which remains a plane of symmetry.

Fig. 34. Distortion of the hydrosodalite lantern toward holohedral symmetry in the structure of Linde sieve A.

The impossibility of a Si—O—Si bond angle close to 180° has a curious effect on the framework of Linde sieve A. This framework is created by superimposing hydrosodalite lanterns, one on the other, the four-membered rings being not shared, but duplicated. From Fig. 26 we can see directly that all four pairs of tetrahedra would have to be directly coupled along edges, while the assumption that such a coupling is impossible is one of the basic tenets of crystal chemistry. So the Chinese lantern shown in Fig. 26, with tetrahedral symmetry ($T_d = 4\bar{3}m$) must be straightened out to the shape shown in Fig. 34 (increasing the symmetry to the maximum cubic, $O_h = m3m$). If we then couple two such tetrahedra, we derive a Si—O—Si angle of 180° which, as we said above, is impossible, so the "straightening out" stops when the compromise angle is reached. As a result, one lantern will not be the exact duplicate of the one underneath it, so the true unit cell of sieve A will not contain one lantern plus one hollow, but 8+8. On the X-ray photograph a large lattice constant (cube edge) of 24 Å is indicated, with a strongly expressed halving, i.e., pseudoconstant of 12 Å. Figure 30 therefore does not show four true unit cells, but only one.

We gave above a clear example of a structural arrangement which was adapted to the shape of a nucleus, which could consist of a foreign substance; this example was the structure of dioptase, $Cu_6[Si_6O_{18}] \cdot 6H_2O$, the nucleus of which is exactly the same as a rhombohedral ring cut out from the structure of ice, the whole structure of the growing crystal being adapted to fit this ring.

Beginning in 1952, some new ideas have been developed on the structure of water, by L. Pauling [45], and after him by H. M. Powell [46] and J. D. Bernal [47]. It appears that there exists a very stable aggregate of 21 particles of H_2O; one in the middle and the other 20 lying around it at the corners of a regular (or almost regular) dodecahedron.

The faces of an ideal dodecahedron, which are regular pentagons, satisfy very well the "natural demands" of the H_2O particle. The face edges have an angle of 108°, very close to the tetrahedral angle of 109°, so that when 20 H_2O particles are placed at the 20 corners of the almost regular dodecahedron, 60 out of the 80 valences of the H_2O, 20aq molecule will lie unconstrainedly at the required angles along the 30* edges of the dodecahedron (in Bernal and Fowler's most widely accepted representation, the H_2O particles have four valences at the "tetrahedral" angle of 109° to each other; two are positive and two negative, each having a value of half a unit; the tetrahedral nature of H_2O is shown particularly clearly in the structure of ice, as may be seen from Fig. 41 of one of the present author's books [18]).

Such dodecahedral molecules of water were found also by Pauling and Marsh [45] in the structure of chlorine hydrate, $Cl_2 \cdot 8H_2O$.† Six formula units, i.e., $6Cl_2 \cdot 48H_2O$, lie in a cubic pseudo-body-centered unit cell of edge 11.88 Å, (which is quite the same as the pseudo unit cell of sieve A with an edge of 12.14 Å, according to [41]). At the $8 + \frac{1}{8}$ corner of the unit cell of the hydrate, and at its center (Fig. 35), we find two large $H_2O \cdot 20aq$ molecules in the form of two dodecahedra‡ rotated by 90°

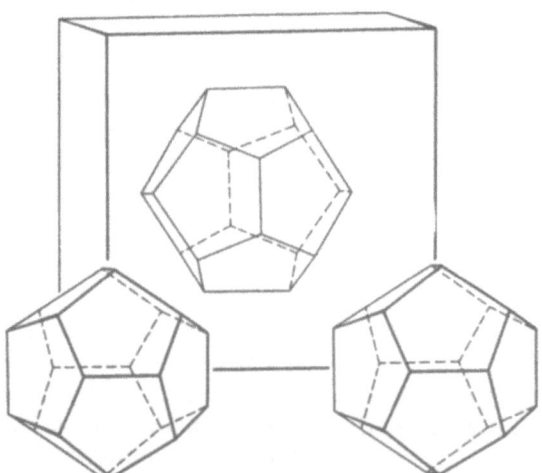

Fig. 35. Structure of chlorine hydrate, $Cl_2 \cdot 8H_2O$. At the corner and in the center of the cube lie dodecahedral water molecules, $H_2O \cdot 20aq$ mutually rotated by 90°. Six further H_2O particles and 12 Cl atoms lie between the dodecahedra (not shown on drawing).

---

*In the regular dodecahedron there are 12 faces, 20 corners, and, according to Euler's theorem, 30 edges.

† They are also found in the hydrates of xenon, bromine, SO_2, H_2S, and others (see Powell's review [46]).

‡ What we have above referred to as a dodecahedral molecule of water should more accurately be called a hydration envelope, and this may be formed, for example, around a CH_4 molecule, a Xe atom, and in particular a 21st molecule of water. It will be appreciated that only the last example can strictly be referred to as a dodecahedral molecule of water.

relative to each other. We stated above that 60 of the hydrogen bonds are saturating each other on the surface of the dodecahedron. Eight more bonds are directed at the necessary angles toward the eight neighboring dodecahedra, along the diagonals of a cube. The twelve remaining bonds are saturated by six H_2O particles which lie between the main dodecahedra ($2 \times 21 + 6 = 48$), together with the six Cl_2 molecules.

In sieve A the large $H_2O \cdot 20aq$ molecule at the center of the cube is retained, but the ones at the corners of the cube are replaced by zeolite cubo-octahedra ("Chinese lanterns"), which are interconnected via the two-storied, four-membered ring to produce a three-dimensional framework. From this geometrical analysis it follows that the formula of sieve A, calculated on one unit cell, must have 21 H_2O particles fewer than that of chlorine hydrate, and in actual fact the formula given for the sieve [39] is $Na_{12}Al_{12}Si_{12}O_{48} \cdot 27H_2O$, containing exactly $48 - 21 = 27$ particles of H_2O, the dodecahedral $H_2O \cdot 20aq$ molecule plus 6 "connecting" H_2O particles. The 12 Cl atoms in the hydrate correspond exactly to the 12 Na atoms in the zeolite.

We noted above that the true unit cell side in sieve A was not 12.14 Å long, but 24.28 Å, and that in consequence the true unit cell contains, not one hollow plus $1(8 \cdot \frac{1}{8})$ cubo-octahedral lantern, but $8 + 8$, so that Fig. 30 shows not four unit cells, but one (if we take into account the changes noted above in the square prism). The edge of the unit cube in sieve X is 24.7 Å, i.e. only a little greater than that in sieve A, and the unit cell also contains eight cubo-octahedra and eight hollows, among which are distributed eight $Na_6Ca_2Al_{10}Si_{14}O_{48} \cdot 30H_2O$ formula units. Thus the amount of water present is only slightly more than that in sieve A, and it is possible (since the water has not, up to the present, been located in sieve X) that in both sieve A and sieve X the "nuclei" are the same, the dodecahedral $H_2O \cdot 20aq$ molecule.

It appears, therefore, that synthesis of both sorts of sieve boils down to very careful handling (low temperature and pressure) of normal $H_2O \cdot 20aq$ molecules. Around this molecule an envelope is first formed of SiO_2 gel and $NaAlO_2$, by the characteristic adaption of the Si tetrahedra to the "tone-setting" nuclei, and then, by normal edge growth, the envelope is built up into a three-dimensional framework. This framework can then lose its nuclei reversibly at temperatures up to 600° without breakdown occurring, the water being completely lost at this temperature to be returned at lower

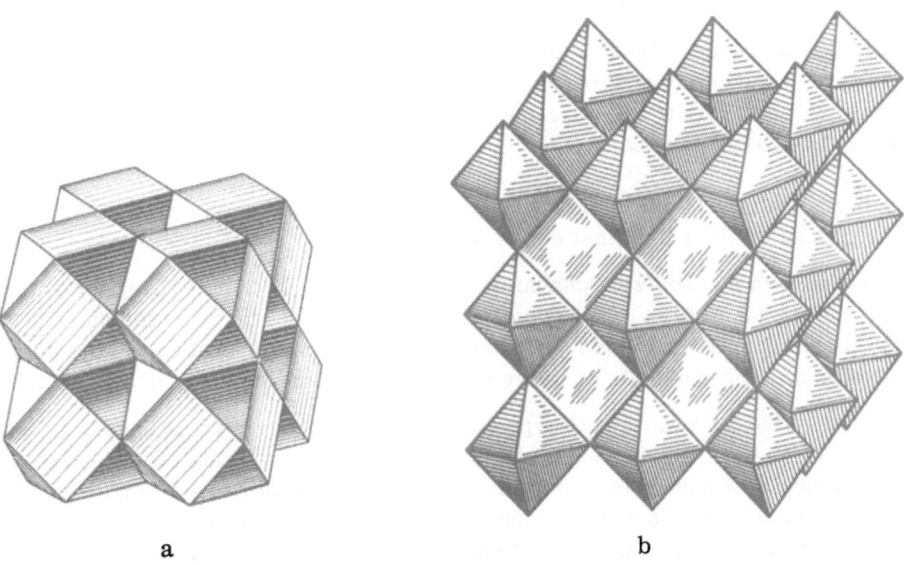

a   b

Fig. 36. Packing of Archimedes cubo-octahedra with an equal number of octa-hedra, as a model for the synthetic zeolite Linde sieve A: (a) Cubo-octahedra only; (b) cubo-octahedra and octahedra.

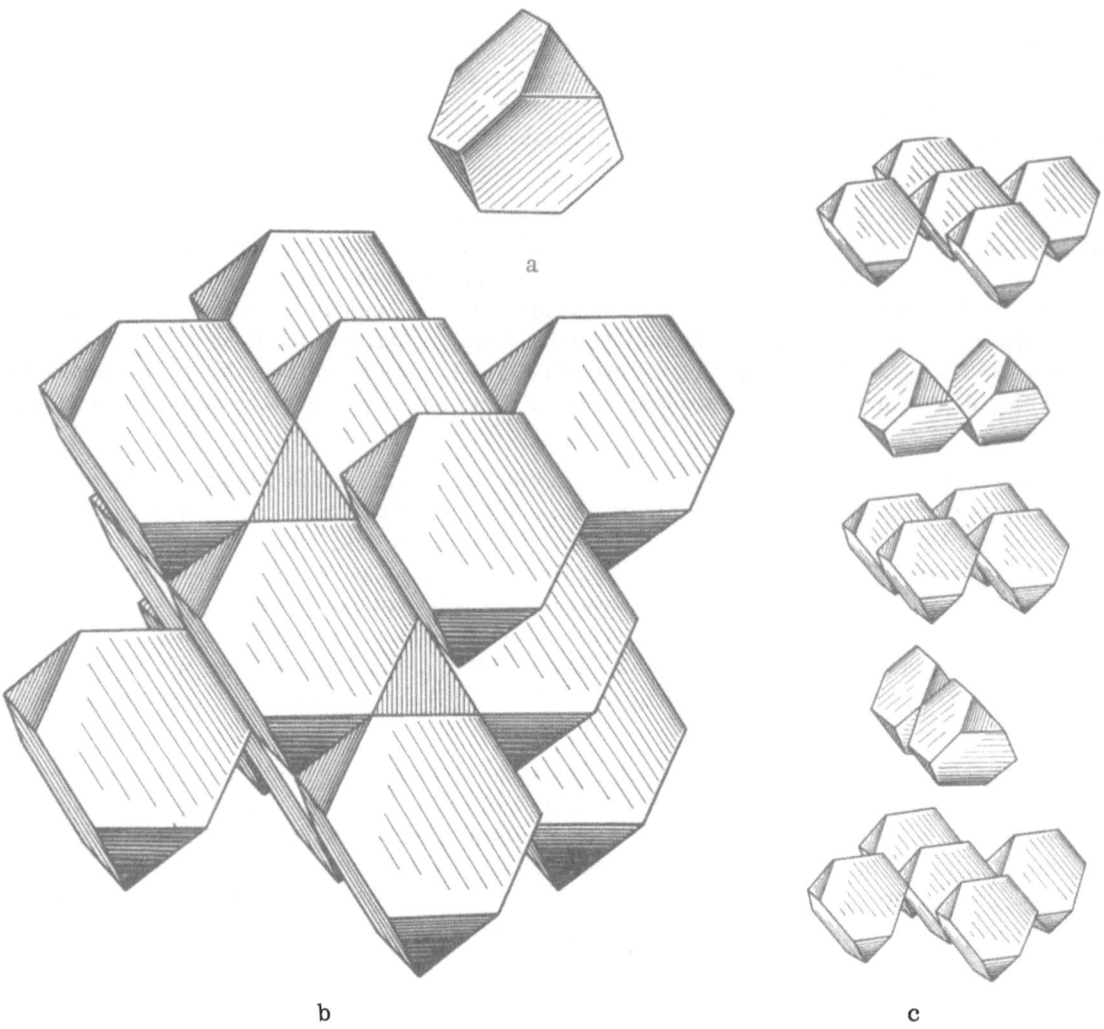

Fig. 37. Packing of Laves polyhedra and an equal number of tetrahedra as a model for Linde sieve X: (a) a Laves polyhedron; (b) face-centered unit cell of Laves polyhedra; (c) the same unit cell split up into the different levels to illustrate the two ("paramorphic") orientations of the Laves polyhedra.

ones or exchanged for other kinds of molecules. Acceptance of other molecules is not governed by their volume, so much as by their smallest cross section.

The lack of success in synthesizing chabazite is now clear. In this zeolite, for every rhombohedral hollow there is one $Na_6Al_6Si_6O_{24} \cdot 12H_2O$ unit, i.e., only half as much water as is contained in one "normal" dodecahedral water molecule. A successful synthesis of chabazite thus depends on a successful preparation of a big molecule of water with half the usual number of H_2O particles.

Thus the volume of the hollows in both types of commercial sieve is the same, although their geometric structures differ; sieve A has six eight-membered windows lying on cube faces, while sieve X has four twelve-membered windows on tetrahedron faces. The differences in constructional principles and, in particular, in methods of interconnecting the hollows, are well illustrated by the simplified models of two kinds of sieves, which we took from [18] (the corresponding symmetry groups are the "primitive" $O_h^1 = Pm3m$ group for sieve A and the "diamond" $O_h^7 = Fd3m$ group for sieve X). Sieve A corresponds to packing of large 12-cornered solids, the Archimedes cubo-octahedra, with an equal number of smaller octahedra (Fig. 36). Sieve X also corresponds to packing of 12-cornered solids but of another type, Laves polyhedra, together with small tetrahedra (Fig. 37). In both arrangements the smaller figures are increased in size at the expense of the tetrahedra in the four- or six-membered rings leading into the hollows, when they build the envelopes of the "lanterns" described above. The hollows in sieve A (Archimedes cubo-octahedra with truncated edges, Fig. 30) themselves possess high symmetry. In sieve X the hollows have tetrahedral form, and the cubic symmetry of the whole is achieved by alternation of the Laves polyhedra in two orientations. (Fig. 37c, polyhedra at different levels are shown separately.)

In the structure of sieve A, the hollows run together to form wide channels in three mutually perpendicular directions, parallel to the {100} directions, and the minimum diameter of these channels is determined by the diameter of the eight-membered ring. It can be seen from Fig. 38a, that sieve X is pierced by channels along the six [110] directions, which are parallel to the six face diagonals of the cubic unit cell. It was emphasized in the explanation to Fig. 33 that the cross section of these last channels is elliptical, even though it is built up from regular 12-sided polygons. In Fig. 38b, we see the simple geometrical explanation of this; the "windows" are inclined alternately, to different sides, along the channel axis. A simple trigonometrical calculation gives a value for the axial ratio of the ellipse of $\sqrt{2/3} = 0.82$.

A recent article [49] on the discovery of a new cubic zeolite, paulingite, with the previously unheard of unit cell side of 35 Å, promises to provide chemical technology with a new, and perhaps particularly powerful, molecular sieve.

A short account in Russian, of the basic technology of these sieves may be found in [50, 51].

* * *

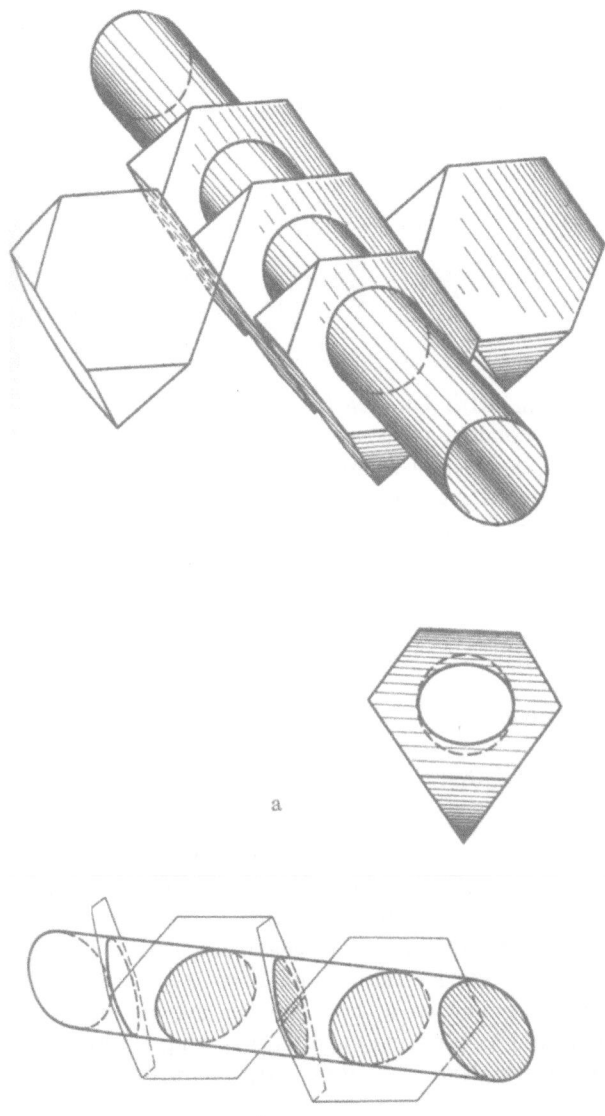

Fig. 38. Channels in sieve X: (a) mode of their penetration through polyhedra; (b) cause of elliptical shape of channel.

In conclusion, we will examine four new silicate structural analyses, carried out in the Moscow laboratories in recent years, on the basis of first and second "chapter" silicate crystal chemistry.

In the structure of epididymite, $NaBeSi_3O_7OH$, yet another type of oxygen-silicon ribbon was discovered [52], which also appears to fall into the second "chapter." This ribbon is produced by condensing two $[Si_{2+1}O_9]_\infty$ wollastonite chains, not through the sharing of every third tetrahedron, as in xonotlite, but through the sharing of all three tetrahedra in each link. Figure 39b,c shows the "two-sided" wollastonite chain, and Fig. 39e, f the products of its condensation, the xonotlite and epididymite ribbons; a minor difference here is that the xonotlite ribbon is produced by operation of a 2-fold axis, while the epididymite ribbon is produced by a mirror plane. The xonotlite ribbon is produced according to the formula

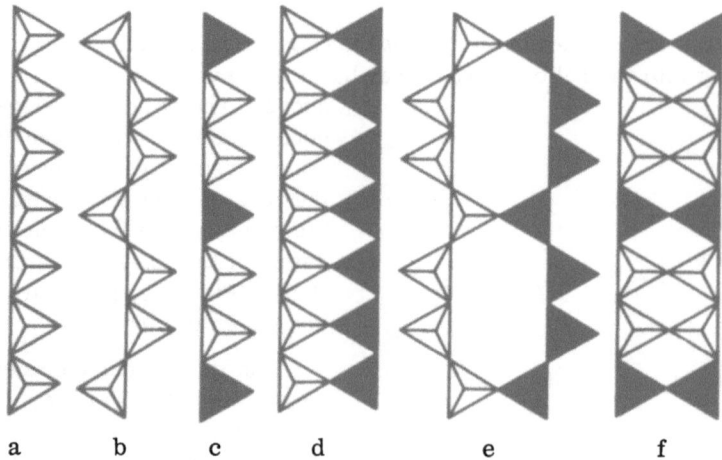

Fig. 39. Oxygen-silicon chains, $[SiO_3]_\infty$, with differently oriented tetrahedra, and some of their condensation ribbons: (a) the simplest $[Si_1O_3]_\infty$ chain; (b), (c) $[Si_{2+1}O_9]_\infty$ wollastonite chains, viewed normal to the plane of centers of gravity of the tetrahedra, and viewed from the side; (d) sillimanite ribbon; (e) xonotlite ribbon; (f) epididymite ribbon.

$$2[Si_{2+1}O_9] - O = [Si_{4+2}O_{17}],$$

while the epididymite ribbon has the following condensation equation:

$$2[Si_{2+1}O_9] - 3O = [Si_{4+2}O_{15}] = [Si_6O_{15}].$$

On dividing this formula by 3, we have $[Si_6O_{15}] = 3[Si_2O_5]$, corresponding to the dimetasilicate formula (O : Si = 2.5 : 1), previously considered as characteristic of the network of platy silicates. As can be seen from Fig. 40, the new type of Si-tetrahedron ribbon arises by adaptation of a dimetasilicate radical to a column of Na cations, which in epididymite are not lying in octahedra, but in twisted cubes (similar to those around Ca in grossular or around Na in the F analog of garnet, cryolithionite $Na_3Al_2Li_3F_{12}$). These twisted cubes (Fig. 40a) lie on a 2-fold screw axis, so that when viewed from any side we have a square face diagonal, parallel to the 2-fold axis of one cube, alternating with the square face edge of the next twisted cube. A Si_2O_7 group is attached to the diagonal, and a SiO_4 group to the edge of the square.

In the doubled formula of epididymite, $Na_2Be_2Si_6O_{14}(OH)_2$, we have only 14 O atoms for every 6 Si atoms, so one of the two OH[*] groups must be included with the silicate radical, which will then have a formula of $[Si_6O_{14}OH]$.

There is no indication from the X-ray data as to which of the 15 O atoms the H atom belongs, so we have to assume it is distributed randomly among the three O atoms shared between Si and Be. This type of replacement by OH, of an O atom in an oxygen-silicon radical, which not so long ago was considered impossible, is now being suggested more and more often; this is mostly among second "chapter" silicates, as for example in the wollastonite chain $[Si_{2+1}O_8OH]$ found in pectolite, $NaCa_2Si_3O_8OH$, [53], which is very close to wollastonite, and in lovozerite, $Na_2Zr[Si_6O_{12}(OH)_6] \cdot 0.5NaOH$ [54].

___

[*]The formula $Na_2Be_2Si_6O_{15} \cdot H_2O$ is not possible, since epididymite loses its water only at very high temperatures ($\sim 800°$).

In the xonotlite ribbon, as in the amphibole ribbon, all three sorts of Si tetrahedra lie almost in the same plane, while in the epididymite ribbon (Fig. 39f) we have four Si atoms at one level and two at another, similar to the conditions in feldspar ribbons. In the latter, however, the formula of the link is $[Si_8O_{20}] = [Si_2O_5]_4 = [Si_{4+4}O_{20}]$, i.e., there are four tetrahedra on each level (Fig. 41). More important, in the feldspars these ribbons only exist as part of a continuous three-dimensional framework, while in epididymite the ribbons are completely discrete.*

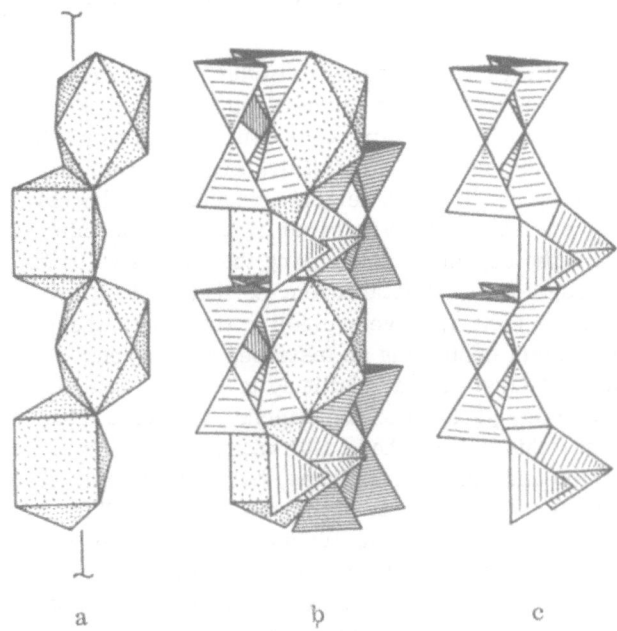

a   b   c

Fig. 40. The two basic components of the structure of epididymite: (a) the column of Na polyhedra; (b) the double chain of Si tetrahedra; (c) the union of both components.

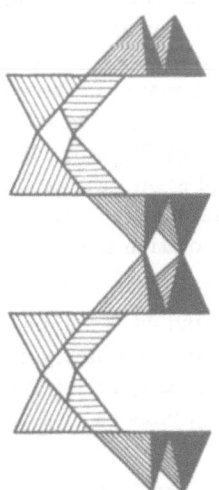

Fig. 41. $[Si_8O_{20}]$ ribbons in the structure of orthoclase.

Ribbons with the link formula $[Si_2O_5]$ are known in the first "chapter" and are depicted in Fig. 39d. Geometrically, they are produced from chains like that shown in Fig. 39a, which are unknown in silicates, and have been found only in copper metagermanate, $CuGeO_3$. Double chains or ribbons of this type are found in sillimanite, Al_2SiO_5, but with the reservation (as in Ca feldspars) that in this ribbon half the tetrahedra are occupied by Al atoms. The length of the link or repeat distance of this chain is only one tetrahedron, so the abbreviated dimetasilicate formula $[Si_2O_5]$ corresponds exactly here to one link (with the reservation that Si is partly replaced by Al).

The basic cause of the formation of second "chapter" silicate structures, made up of $[Si_2O_7]$ units, is the presence in a structure of "indivisible spans" of ~ 4 Å. This span is not necessarily based on the edge of the octahedron or twisted cube around a large cation. We showed above (p. 13) how the Ti octahedra found in a certain arrangement with Si_2O_7 groups were, to a large extent, Ti^{3+} octahedra. The more usual Ti^{4+} octahedron, which is relatively small, appears mostly in first "chapter" silicates, on an equal footing with Mg(Fe) octahedra. In pyroxenes and amphiboles, Ti to a large extent simply replaces Mg(Fe), in octahedra which are connected by shared edges; in ramsayite [18] in a completely analogous ribbon, we have Ti octahedra only, joined by shared edges, to which are attached (in accordance with first "chapter" principles) pyroxene chains of formula $[Si_2O_6]\infty = [Si_{1+1}O_6]\infty$. The geometrical equivalence of Ti and Mg octahedra, shown very clearly by the identical structures of rutile, TiO_2, and sellaite, MgF_2, is demonstrated even more effectively in the norbergite-chondrodite-humite-clinohumite series, which has the

---

*As stated on p. 14, the same epididymite ribbon $[Si_6O_{15}]\infty$ has been found in the structure of elpidite $Na_2Zr[Si_6O_{15}]3H_2O$.

general formula $n\text{Mg}_2\text{SiO}_4 \cdot \text{MgF}_2$. I. D. Borneman-Starynkevich has shown effectively that in these minerals, the MgF_2 group may be completely replaced by TiO_2 [55].

Many minerals have a large amount of SiO_2 combined with a small number of Ti^{4+} tetrahedra, and every oxygen atom forming the octahedron around Ti is also part of a tetrahedron around Si (oxygen starvation!). In these cases it is unlikely that neighboring Ti octahedra share edges* (as in rutile and ramsayite). When the O:Ti ratio is small, that is, when isolated Ti octahedra are no longer possible, the octahedra join up into chains with the AX_5 motif [18], in which neighboring Ti octahedra have only one shared O(OH) corner (Fig. 42). In this chain the repeat distance along the axis is $\sqrt{2}$ times the edge of a Ti^{4+} octahedron, so that its length, $2.85\sqrt{2} = 4.00$ Å, is very close to the length of edge of a prism containing an inscribed Si_2O_7 group. When Si_2O_7 groups are built up into an oxygen-silicon chain, parallel to and attached to such a TiO_5 chain, this will not have alternating Si_2O_7 diortho groups and SiO_4 ortho groups, but will contain links made up entirely of diortho groups. This gives a chain containing links of one type only, with the empirical formula $[\text{SiO}_3]_\infty$ and the structural formula $[\text{Si}_{2+2}\text{O}_{12}]_\infty$. Chains of this type have been found in the tetragonal mineral narsarsukite, $\text{Na}_2(\text{Ti, Fe})(\text{O, OH})[\text{Si}_4\text{O}_{10}]$ [56]. The links are joined into a chain which runs down a 4-fold screw axis (Fig. 43), and the chain is paired up with another, similar, chain which forms a second turn on the same screw axis; both chains together form a double-turn 4-fold screw axis, 4_2. This is shown in Fig. 43, which also shows that every tetrahedron in the chain participates in the pairing, as in epididymite and sillimanite. The ribbon will therefore have the formula $[\text{Si}_2\text{O}_5]_{4\infty}$, or more precisely, $[\text{Si}_{4+4}\text{O}_{20}]_\infty$. Figure 44 shows the complete plan of the structure of narsarsukite, with $[\text{TiO}_5]$ chains passing along 4-fold rotation axes and $[\text{Si}_8\text{O}_{20}]_\infty$ ribbons winding around 4_2 4-fold screw axes.†

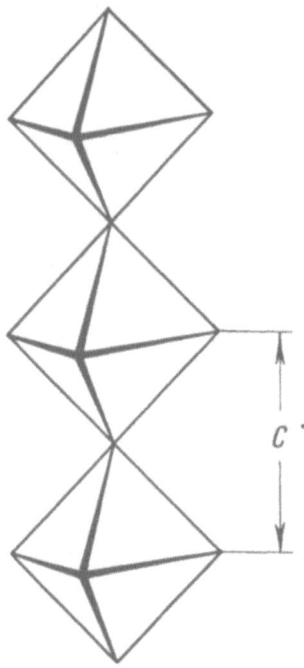

Fig. 42. The AX_5 motif, consisting of octahedra in which the repeat distance along the chain axis (C) is equal to the diagonal of the octahedron—the diagonal of a square constructed on an edge of the octahedron.

From Figs. 43 and 44 it can be seen that the oxygen-silicon ribbon can also be split up into four-membered rings with the symmetry of 4-fold inversion axes, $\overline{4}$. Two tetrahedra in each ring point upward, and two downward, forming an infinite ribbon with tetragonal symmetry. Between the TiO_5 chains and the $[\text{Si}_8\text{O}_{20}]$ ribbons, Na cations lie in hexagonal channels, in an arrangement recalling that in feldspars.

Unlike $\text{Na}_2\text{TiO}[\text{Si}_4\text{O}_{10}]$, the mineral baotite, $\text{Ba}_4(\text{Ti, Nb})_8\text{O}_{16}[\text{Si}_4\text{O}_{12}]\text{Cl} - $ [57], also tetragonal, has a lot more titanium (partially replaced by niobium) than silicon, and so the Ti octahedra must be joined up along their edges, as in the three TiO_2 modifications and in ramsayite, creating conditions which place the mineral in the first "chapter." Figure 45 shows how columns of Ti octahedra, connected by shared horizontal edges, stand at the centers of the four quadrants of the tetragonal unit cell in an arrangement which is simply cut out of the structure of tetragonal rutile [18], with Ti octahedra

---

*Otherwise one O anion would be shared between two Ti octahedra and one Si tetrahedron, which would contravene Pauling's second rule.

†Note that on p. 15 it is stated that pure chains with only $[\text{Si}_2\text{O}_2]$ links have been found in batisite $\text{Na}_2\text{BaTiO}_2[\text{Si}_4\text{O}_{12}]$.

40

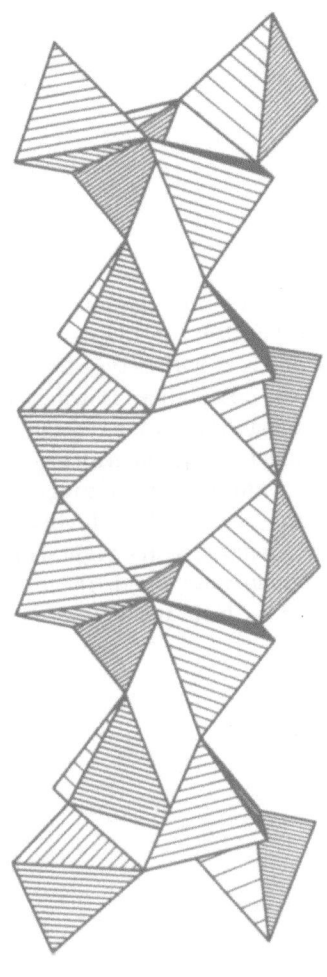

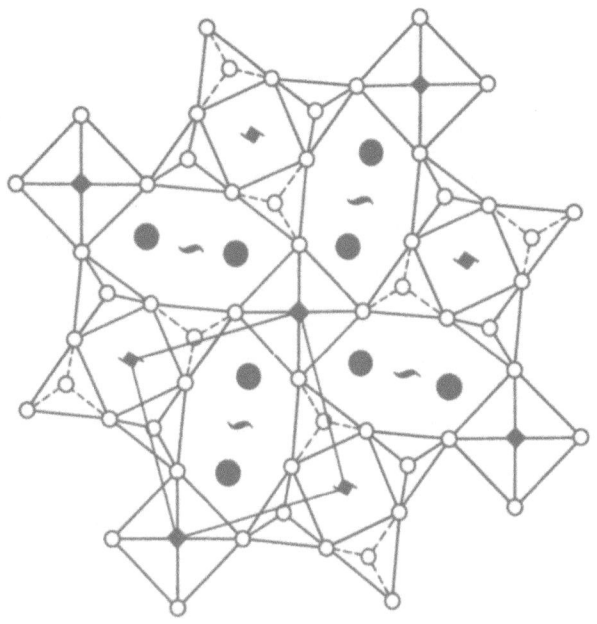

Fig. 43. The tubular narsarsukite oxygen-silicon ribbon with the dimetasilicate formula, $[Si_8O_{20}] = [Si_2O_5]_4$. It is a model for a double-turn 4-fold screw axis 4_2 with perpendicular symmetry planes $(4_2/m)$.

Fig. 44. The tetragonal narsarsukite structure (plan). Chains of Ti octahedra down the side edges of the unit cell, and down the central axis, alternate with tubular ribbons of Si tetrahedra. The structure is given in the base-centered representation to show up the channels containing Na atoms which pass down 2_1 along the 2-fold screw axes.

lying in pairs at two levels (4_2 axis). At the four corners of each quadrant we find a four-membered $[Si_4O_{12}]$ ring, which as this is a first "chapter" silicate, is only single-storied. The discovery of this four-membered ring in baotite, by V. I. Simonov [57], filled an annoying gap in the systematic classification of silicate radicals.*

In each baotite quadrant, the $[Si_4O_{12}]$ metasilicate rings are arranged around a central rutile motif at four levels, corresponding to a single-turn 4-fold screw axis, right-handed (4_1) in quadrants I and III, left-handed ($4_{-1} = 4_3$) in II and IV. Between the vertically repeated four-membered rings lie large Cl anions, in an arrangement recalling that in sodalite [18]. In particular, where in sodalite

---

*The $[Si_4O_{12}]$ ring was proposed by T. Ito [58] for the borosilicate axinite, but without the reliable proof which is necessary for such a complex structure. A note on the possible existence of this ring in neptunite, $Na_2FeTi[Si_4O_{12}]$, which appears in Strunz's tables [49] as his personal view, remains an unconfirmed guess, of doubtful validity in view of the closeness of neptunite to pyroxenes. However, analogous $[P_4O_{12}]$ rings were found a long time ago [60] in the cubic metaphosphate $Al[PO_3]_3 = Al_4[P_4O_{12}]_3$. In 1962 we established this $[Si_4O_{12}]$ group in the structure of kainosite (G. F. Volodina et al. Doklady AN SSSR, 149, No. 1, 1963).

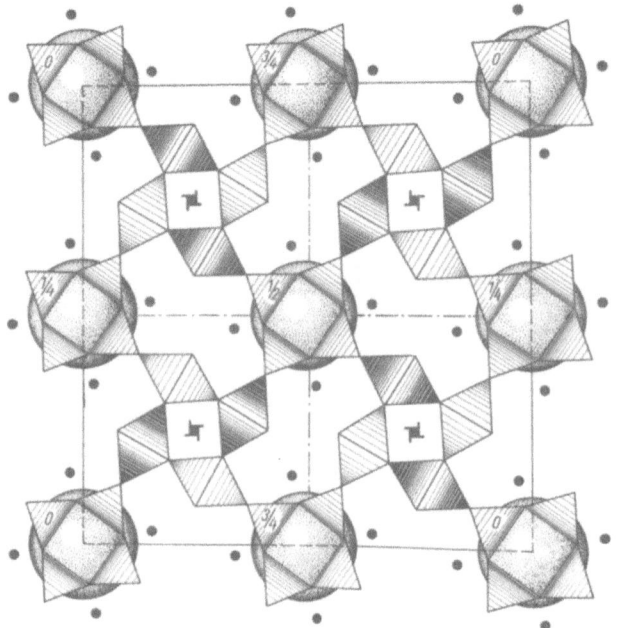

Fig. 45. Plan of the tetragonal structure of baotite. In the center of each quadrant is a rutile-type arrangement of Ti(Nb) octahedra. Down the outside edges of each quadrant, isolated four-membered Si_4O_{12} rings alternate with large Cl atoms. These rings, and the Cl atoms, rest at four levels around screw axes; these are right-handed (4_1) in quadrants I and III, left handed ($4_{-1} = 4_3$) in II and IV.

Fig. 46. Pseudocubic cells in the structure of lovozerite. Zr atoms in octahedra occupy the cell corners; eight-cornered polyhedra around Na form the edges. Inside each cell is a six-membered $[Si_6O_{18}]$ ring with two vertical Si_2O_7 groups (normal to plane of drawing), connected by two Si tetrahedra at different levels.

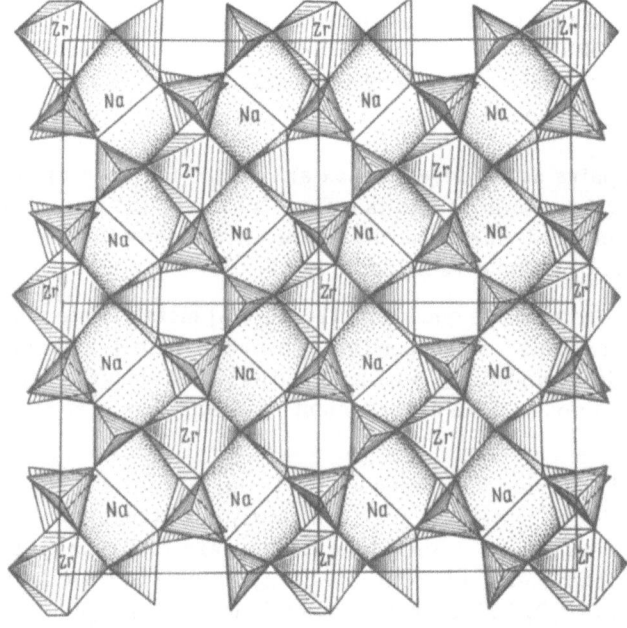

each Cl atom is surrounded by eight Na cations disposed in six-membered "Chinese lantern" rings, in baotite each Cl atom is surrounded by four Ba cations, which are too large to satisfy the requirements of the second "chapter." The same applies to the large Cl anion, but we must take into account the fact that both are few in number compared to Ti atoms, which, as we have said, determine that baotite falls into the first "chapter" with single-story oxygen-silicon rings.

Classifying silicates as first or second "chapter" is a useful method for remembering a series of properties of minerals and synthetic compounds. However, the classification must not be assumed to be inviolate. We note, for example, that the fundamental second "chapter" chain, the $[Si_3O_9]_\infty$ metasilicate wollastonite chain, is made up of alternating $[Si_2O_7]$ and $[SiO_4]$ groups, i.e., the second "chapter" building unit is interspersed with first "chapter" units. Pure second "chapter" metasilicate chains (like those in narsarsukite*) are extremely rare. In the epidote-zoisite group the Si_2O_7 unit is the building brick in the y-axis direction, where it is matched with Ca polyhedron ribbons and the SiO_4 unit is the brick in the z-axis direction, parallel to columns of Al octahedra. In ilvaite we have the unusual situation of $[Si_2O_7]$ groups behaving as first "chapter" units; they form a kind of oxygen-silicon "double ring," the analog of the single-storied three, four, and six-membered rings which, in a previous section, we somewhat carelessly referred to as first "chapter" radicals, but which in the following case will appear also as second "chapter" units.

The structure of lovozerite, $Na_2ZrSi_6O_{15}$ $\cdot 3H_2O \cdot 0.5 NaOH$, really does have a lattice-like appearance [54]. It consists of "cubic" cells (Fig. 46) in which the cell side is 7.31 A, but the roof of each cell is shifted diagonally relative to its base, giving an angle β not of 90°, but of 92°30'. At the corners of each cell are Zr atoms in oxygen octahedra, and at the middle of all twelve edges are Na cations, lying in drawn-out polyhedra which are a type of deformed cube, best described as a pair of trigonal prisms joined together on a square face.

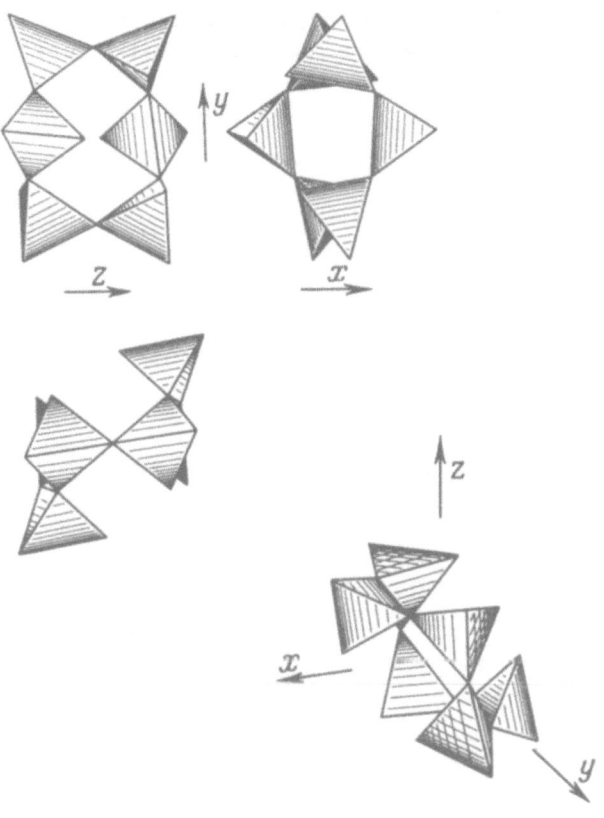

Fig. 47. The six-membered $[Si_6O_{18}]$ ring in the structure of lovozerite—projections along the three axes, and a perspective drawing. A horizontal 2-fold axis passes through the bridge O atoms of the two Si_2O_7 groups lying parallel to the z-axis.

The individual cells contain "island" radicals of formula $[Si_6O_{18}]$, and this metasilicate group is actually a six-membered ring. Although in beryl and dioptase the six-membered rings have high symmetry, 6/m in the first case and $\overset{\circ}{6} = \overline{3}$ (6-fold mirror axis) in the second, the ring in lovozerite has only a 2-fold axis which lies in the plane of the ring, while the ring is inclined to the x and z axes at about 45°. When examined closely, it is found that the lovozerite ring can also be represented as two $[Si_2O_7]$ diortho groups running along the z-axis and joined to "vertical" Na polyhedra according to second "chapter" principles. The two parallel Si_2O_7 groups are attached to diagonally opposite corners of the cell by single Si tetrahedra, one above and one below (related by a 2-fold axis), as a result of which a six-membered ring is formed, within which four further horizontal Si_2O_7 groups can be distinguished, attached to the other, "horizontal," Na polyhedra.

---

*Or in batisite (p. 15).

Thus the fundamental structural feature of this silicate is the adaptation of the oxygen-silicon radical to second "chapter" conditions created by a rigid cellular lattice of Zr and Na polyhedra.

According to the formula given above for lovozerite, all the water of constitution must be included in the oxygen-silicon radical in the form of OH groups, giving a formula for the ring of $[Si_6O_{12}(OH)_6]$. As an example of the introduction of the OH group into the oxygen-silicon radical (p. 38), this is a most interesting case, and may possibly be connected with the poor natural formation and instability of lovozerite. The existence of only half a molecule of NaOH in the "side chain" has an interesting explanation. X-ray results show that only half the vertical edges of the cell are occupied by Na atoms. We must assume that the other half are occupied either by free hydroxyl groups, or, which is more likely, by oxonium groups, OH_3^+, which have the same valence as Na. Such a replacement would lower the number of OH groups replacing O in the oxygen-silicon radical.

Literature Cited

[For the reader's convenience, some of the cited Russian literature available in English translation has been reprinted in the appendix. Such entries are marked with asterisks (*), and their locations are indicated in brackets.]

1. B. K. Brunovskii. Acta phys. chem. USSR, 5, 863 (1936).
2. N. V. Belov, V. P. Butuzov, and N. I. Golovastnikov. Dokl. AN SSSR, 37, 156 (1942); 87, No. 6, 953 (1962).
3. N. V. Belov and L. M. Belyaev. Dokl. AN SSSR, 69, No. 6. 805 (1949).
4. N. V. Belov and T. N. Tarkhova. Dokl. AN SSSR, 69, No. 3, 385 (1949); Tr. In-ta Kristallogr. 6, 25 (1949).
5. N. V. Belov and V. I. Mokeeva. Dokl. AN SSSR, 81, No. 4. 581 (1951); Tr. In-ta Kristallogr. 9, 49 (1954).
6. N. V. Belov and I. M. Rumanova. Dokl. AN SSSR, 89, No. 5, 853 (1953); Tr. In-ta Kristallogr. 9, 105 (1954).
7. E. G. Fesenko, I. M. Rumanova, and N. V. Belov. Dokl. AN SSSR, 102, No. 2. 275 (1955); Kristallografiya, 1, No. 2, 171 (1956).
8. R. F. Smirnova, I. M. Rumanova, and N. V. Belov. Zap. Vses. Mineralog. ob-va, pt. 84, 2, 159 (1955).
9. Kh. S. Mamedov and N. V. Belov. Dokl. AN SSSR, 104, No. 4, 615 (1955); Zap. Vses. Mineralog. ob-va, pt. 85, 1 (1956).
10. Kh. S. Mamedov and N. V. Belov. Dokl. AN SSSR, 107, No. 3, 463 (1956).
*11. P. V. Pavlov and N. V. Belov. Dokl. AN SSSR, 114, No. 4, 884 (1957); Kristallografiya, 4, No. 3, 324 (1959). [See Appendix p. 82.]
*12. V. I. Simonov and N. V. Belov. Dokl. AN SSSR, 122, No. 3, 473 (1958); Kristallografiya, 4, No. 2, 163 (1959). [See Appendix p. 67.]
*13. V. I. Simonov and N. V. Belov. Dokl. AN SSSR, 130, No. 6, 1333 (1960)[See Appendix p. 97.]; Silikat. Tech., 11, 8, 392 (1960).
*14. V. V. Ilyukhin and N. V. Belov. Dokl. AN SSSR, 131, No. 1, 176 (1960); Kristallografiya, 5, No. 2, 200 (1960). [See Appendix p. 101.]
*15. E. A. Pobedimskaya and N. V. Belov. Dokl. AN SSSR, 129, No. 4, 900 (1959); Zhurn. strukt. khim., 1, 1, 51 (1960). [See Appendix p. 114.]
16. Kh. S. Mamedov. Dokl. AN Azerb.SSR, 14, No. 6, 445 (1958).
*17. Kh. S. Mamedov and N. V. Belov. Dokl. AN SSSR, 123, No. 4, 741 (1958). [See Appendix p. 60.]
18. N. V. Belov. Structures of Ionic Crystals and Metallic Phases [in Russian]. Izd-vo AN SSSR, 1947.
19. N. V. Belov. Essays on Structural Mineralogy. X. Mineral. sborn. L'vovskogo geol. ob-va, No. 13, 23 (1959).
20. J. V. Smith. Acta Crystal., 6, 9 (1953).
*21. Kh. S. Mamedov, R. F. Klevtsova, and N. V. Belov. Dokl. AN SSSR, 126, No. 1, 151 (1959). [See Appendix p. 79.]
22. E. I. Semenov, M. E. Kazakova, and V. I. Simonov. Zap. Vses. mineralog. ob-va, pt. 87, 5 (1958).
23. K. A. Vlasov, M. V. Kuz'menko, and E. M. Es'kova. The Lovozero Alkaline Massif [in Russian]. M, Izd-vo AN SSSR, 1959.
24. A. I. Tsvetkov. Tr. In-ta geol. Nauk AN SSSR, ser. petrograf., 138, 41 (1951).
25. V. S. Sobolev. Introduction to Silicate Mineralogy [in Russian]. L'vov, 1947.
26. N. V. Belov. Acta Crystal., 10, 757 (1957).

*27. Kh. S. Mamedov and N. V. Belov. Dokl. AN SSSR, 121, No. 5, 901 (1958). [See Appendix p. 57.]

28. J. D. Bernal and H. D. Megaw. Proc. Roy. Soc. London, 151A, 384 (1953).

29. L. Heller and H. F. W. Taylor. Crystallographic Data for Calcium Silicates. London, 1956.

30. F. Liebau. Acta Crystal., 10, 790 (1957).

31. F. Liebau, M. Sprung, and E. Thilo. Z. Anorg. Chemie, 297, 263 (1958).

32. W. Hilmes, F. Liebau, and E. Thilo. Die Naturwissenschaften, 43, 177 (1956).

33. F. Liebau. Die Naturwissenschaften, 44, 6, 178 (1957); Acta Crystal., 12, 177 (1959).

34. L. S. Dent-Glasser and F. P. Glasser. Silikat. Tech., 11, 8, 363 (1960).

*35. Kh. S. Mamedov and N. V. Belov. Dokl. AN SSSR, 121, No. 4, 720 (1958). [See Appendix p. 53.]

36. Kh. S. Mamedov and N. V. Belov. Dokl. AN SSSR. 123, No. 1, 163 (1958).

37. N. V. Belov. Essays on Structural Mineralogy. XI. Mineral. sborn. L'vovskogo geol. ob-va, No. 14, 3 (1960).

38. L. S. Dent and J. V. Smith. Nature, 181, No. 4626, 1794 (1958); J. V. Smith. Abstracts of Papers for the meeting of the Am. Min. Soc., 1960, 212.

39. D. W. Breck, W. G. Eversole, R. M. Milton, T. B. Reed, and T. L. Thomas. J. Am. Chem. Soc., 78, 5063 (1956).

40. R. M. Barrer and W. M. Meier. Trans. Faraday Soc., 54, N. 427, 1074 (1958).

41. T. B. Reed and D. W. Breck. J. Am. Chem. Soc., 78, 5972 (1956); P. A. Howell. Acta Crystal., 13, 737 (1960).

42. R. M. Barrer et al. Proc. Roy. Soc., 237A, 439 (1956); Helv. Chim. Acta 39, 518 (1956); Trans. Faraday Soc. 52, 1111 (1957).

43. G. Bergerhoff, W. H. Baur, and W. Nowacki. N. Jahrbuch Mineral., 9, 193 (1958).

44. R. M. Barrer. New Selective Sorbents. Brit. Chem. Eng., 1959. p. 1-13.

45. L. S. Pauling, R. E. Marsh. Proc. Nat. Acad. Sci., U.S., 36, 112 (1952): L. Pauling. The Nature of the Chemical Bond. 3rd edition, 1960.

46. H. M. Powell. J. Chem. Soc., 1948, 61; Research (London), 1, 353 (1948); J. Chem. Soc., 1950, 298, 468.

47. J. D. Bernal. Nature, 183, 141 (1959).

48. J. D. Bernal and R. H. Fowler. J. Chem. Phys., 1, 515 (1933).

49. W. B. Kamb and W. C. Oke. Amer. Mineral., 45, 78 (1960).

50. N. V. Kel'tsev. Gazovaya promyshlennost', No. 9, 38 (1957).

51. G. V. Tsitsishvili and G. S. Andronikashvili. Molecular Sieves [in Russian] Tbilisi, Izd-vo Gruz. AN. 1960.

*52. E. A. Pobedimskaya and N. V. Belov. Dokl. AN SSSR. 104, No. 4 (1959); Zhurn. strukt. khim., 1, 1, 51 (1960). [See Appendix p. 114.]

53. M. J. Buerger. Z. Kristal., 108, 248 (1956).

54. V. V. Ilyukhin and N. V. Belov. Dokl. AN SSSR. 131, No. 1, 176 (1960); Kristallografiya, 5, No. 2 (1960).

55. I. D. Borneman-Starynkevich and V. S. Myasnikov. Dokl. AN SSSR, 71, No. 1, 137 (1950).

*56. Yu. A. Pyatenko and Z. V. Pudovkina. Kristallografiya, 4, No. 6 (1959); 5, No. 4, 563 (1960). [See Appendix p. 128.]

*57. V. I. Simonov. Kristallografiya, 5, No. 4, 544 (1960). [See Appendix p. 125.]

58. T. Ito and J. Takéuchi. Acta Crystal., 5, 202 (1952).

59. H. Strunz. Mineralogische Tabellen. 3rd edition. Leipzig, 1957.

60. L. Pauling and J. Sherman. Z. Kristal., 97, 481 (1937).

APPENDIX

Reprinted from Soviet Physics—Crystallography, Vol. 6, pp. 661-664, May-June, 1962

NIKOLAI VASIL'EVICH BELOV

(On His Seventieth Birthday)

Translated from Kristallografiya, Vol. 6, No. 6,
pp. 815-819, November-December, 1961

Academician Nikolai Vasil'evich Belov was born on December 15, 1891 in the town of Janow of the former province of Lubinsk (Poland) into the family of a doctor. He finished the Warsaw High School in 1910 with a gold medal and entered the Metallurgical Department of the St. Petersburg Polytechnic Institute. Among the galaxy of famous professors who in those years were building up the reputation of the Polytechnic Institute as the leading technical college of its time —N. S. Kurnakov, V. A. Kistyakovskii, F. Yu. Levinson-Lessing, A. A. Baikov, M. A. Pavlov, P. P. Fedot'ev and others—N. V. Belov came under the particular influence of the young A. F. Ioffe.

Even while he was at the Institute, and independently afterwards, N. V. Belov eagerly absorbed the remarkable achievements of physics and chemistry—leading sciences forming two of the three corners of the "triangle," the middle of which was accepted as the position of crystallography. The mastering of the third corner, mineralogy, and the unique synthesis of all three of them in the scientific work of N. V. Belov occurred some time later.

In the first few years after the revolution, N. V. Belov devoted considerable effort to the rehabilitation of the country's industry, and in 1920-1921 he was president of the District Council of People's Commissars in Ovruch. In the 20's he worked in analytical chemical laboratories in Leningrad and on the academic journal "Priroda," where he came into close contact with A. E. Fersman. The latter encouraged him to work on the then central problem of Khibiny: N. V. Belov suggested a number of valuable uses for nepheline in the leather and other branches of light industry. The desire to discover the physicochemical significance of the remarkable properties of this silicate led N. V. Belov to the only path along which this could be attained—to microcrystallography. It appeared, however, that the decisive step on this path, which rapidly made N. V. Belov an authority on structural crystallography, was the translation into Russian, made at the suggestion of A. E. Fersman in 1934, of O. Hassel's "Crystal Chemistry," which summed up and generalized the principal achievements in this science up to that time. It is significant, however, that the creative qualities of N. V. Belov could not be reconciled with such a routine process as scientific translation. The result was that the volume of the Russian text was twice the size of the German version, the number of illustrations increased from 6 to 60, and the translator of the Russian version became virtually the joint author. At this time (1933), N. V. Belov was employed on a permanent basis by the Academy of Sciences, USSR (in the Lomonosov Institute of the Academy).

What were the achievements of structural crystallography at that time?. The indisputable authorities and leaders of x-ray crystallographers were W. H. Bragg and his son, the still-thriving W. L. Bragg, who, with pupils such as L. Pauling, W. H. Zachariasen, B. Warren, W. H. Taylor, G. West, S. Naray-Szabo, F. Machatschki, and T. Ito, interpreted the structure of many of the most complicated of atomic configurations of inanimate nature, the silicates. Chemists and mineralogists, who for a century and a half had hesitatingly faced the problem of the structure of silicates with the means available to them, saw for themselves the power of the new physical method of x-ray analysis, which, while solving the problem of chemical structure, provided something more valuable still, i.e., the crystal-chemical structure of matter, the mutual arrangement of the atoms, the length of the bonds and the angle between them, and everything else which solves the problem of the nature of the chemical bond. In volume III of "Fundamental Ideas of Geochemistry," published in 1935, N. V. Belov assembled all the principal work on structure of silicates.

Unfortunately, following the fundamental work of E. S. Fedorov and G. V. Wulff, the 1920's and the beginning of the 1930's were not very fruitful years for Russian microcrystallography, apart from some outstanding work by x-ray metallurgists. N. V. Belov was destined to become a leading figure first in the development of the Soviet school of thought on structure and later in its prime. The commencement of this fruitful period of N. V. Belov's activities is associated with his transfer in 1936, after the Academy had moved to Moscow, to the Crystallographic Department of the Lomonosov Institute headed by A. V. Shubnikov and later becoming the Institute of Crystallography, Academy of Sciences, USSR. From 1938 to the present time, N. V.

Belov has been the permanent head of the Structure Department of the Institute.

A characteristic and obviously most valuable quality of an outstanding scientist is his scientific intuition, enabling him in the development of a science to feel the location of the branch which, at a given stage, appears to hold out the greatest promise and to be capable of making the most rapid progress, and in what direction his efforts and those of his pupils must concentrate in making the principal attack. N. V. Belov possesses this quality to the fullest degree.

The concept of the close packing of the particles forming a crystal structure permeates the whole development of crystallography, both in the period before x-ray analysis and in the current x-ray period. At the end of the 1930's and the commencement of the 1940's, when the classical method of "trial and error" in the interpretation of complex crystallographic structures began to result more frequently in misses, N. V. Belov developed and transformed the theory of close packing and coordination polyhedra, converting it not only into a perfect and elegant method of describing structures, but also into a powerful method of interpreting them. N. V. Belov deduced all the possible methods of packing spheres in given layers and all the space groups characterizing them (and, as it appeared, there are eight of these altogether, i.e., those containing P3m as subgroup). Having constructed new types of coordination polyhedra on this basis, he considerably reduced the number of structural modifications which have to be examined by the trial method. In fact, it was in this way that N. V. Belov and his pupils L. M. Belyaev, V. P. Butuzov, T. N. Takhova, E. N. Belova, V. I. Mokeeva and others in the 1940's interpreted the very complex structure of such silicates as ramsayite, dioptase, tourmaline, ilvaite, milarite, and others. N. V. Belov gave a generalization of the theory of closest packing in his well-known book "Structure of Ionic Crystals and Metal Phases" (1947) containing a large number of original models and motifs, now constantly reproduced in Russian and foreign publications.

In an endeavor to simplify and shorten laborious structural analyses, N. V. Belov at that time produced a number of simple and effective methods of computation, and developed and introduced into all the x-ray structure laboratories of the country the then most advanced method of strip summation of Fourier series, which only recently has been replaced by the use of electronic computers. N. V. Belov's nomogram methods still remain among the basic methods in the initial stages of structural analysis.

The whole of crystallography, both macro- and microcrystallography is based on a study of symmetry. The remarkable legacy of E. S. Fedorov — the theory of space groups of symmetry —has found its finest and most profound interpreter in the person of N. V. Belov. Having devised a new and remarkably simple algorithm for the deduction of Bravais lattices and Fedorov groups (the "class" method), revealing clearly the connection between each of them and the others and the original point group, N. V. Belov made available to every student a clear understanding of their geometrical properties and, what is no less important, their diffraction properties (deduction of formulas for the structural factors), whereas formerly such understanding was the privilege of not more than a dozen chosen specialists throughout the entire world. By making use of this algorithm, it was found possible to deduce not only the 230 Fedorov-Schoenflies groups, which had long been tabulated, but also the previously unknown 1651 groups of space antisymmetry. This problem was solved by N. V. Belov and his pupils N. N. Neronova and T. S. Smirnova in 1954, with an ease which is astonishing in comparison with the classical but very unwieldly derivation by A. M. Zamorzaev, who, at the suggestion of A. V. Shubnikov in Leningrad, developed the concepts of black-and-white (anti)symmetry. The antisymmetry space groups were found suitable for describing the ferromagnetic and electrical properties of crystals, and a corresponding selection of them was made by N. V. Belov and his pupils. Under the influence of N. V. Belov, the name of E. S. Fedorov was given its rightful prominent place in the second edition of the "International Tables." N. V. Belov was the instigator of the Internation Session on Crystallography held in Leningrad 1959 to commemorate the great Russian scientist.

The most important matter in antisymmetry is to assign opposite (plus and minus) properties to geometrically identical (compatible or mirror) images. One may go further and, with N. V. Belov, increase the number of properties transformed from one to the other from two to three, four or six. We then get the Belov color symmetry. It is embodied not only in picturesque mosaics, illustrating each Belov group of color symmetry, but it has recently become clear that the operation of color symmetry can be applied to the examination of the properties of Fourier transforms in inverse space. There is no doubt that color symmetry will find other applications.

As long ago as in the structural work of the 1940's, a remarkable series of papers on the direct methods of the x-ray analysis of crystals began to be published by N. V. Belov and his pupils. The unit cells of the crystals investigated became larger and larger, and the number of parameters describing the position of atoms in them also increased. Methods were required for deriving the structure directly from experimental data, independently of any crystal-chemical assumptions, i.e., direct methods. On the two general lines of direct methods — derivation of structure from the Patterson function of interatomic distances, and the direct determination of the signs of the structure amplitudes — the Belov school ob-

tained important results, embodied in quite a number of new structure determinations. The practical methods devised were put into permanent general use. These were theorems on the interpretation of the F^2 series of hexagonal crystals, determination of the center of symmetry from F^2 series, development of minimalizing methods, development of a statistical method of determining the signs, taking microsymmetry into account, development of the method on inequalities and so forth. They were used in the 1950's for interpreting the structure of the following: epidote, zoisite, eremeyevite, cuspidine, xonotlite, wollastonite, chkalovite, herderite, gadolinite, herlbutite, amblygonite, beryllonite, seidozerite, lovenite, epididymite, eudidymite, lovozerite, bertrandite, baotite and many others (this work was done under the leadership of Belov by I. M. Rumanova, Kh. S. Mamedov, N. I. Golovastikov, V. I. Simonov, E. A. Pobedimskaya, V. V. Bakanin, E. G. Fesenko, V. V. Ilyukhin and others).

During these investigations, it gradually became clear that the systematics of silicates, developed by W. L. Bragg, L. Pauling, E. Schiebold, F. Machatschki and other scientists, and for about 30 years passing from one textbook to another, was far from being complete. It included many silicates with the cations Mg, Al, Fe in octahedra, the edges of which were nearly the edges of the fundamental block of the silicate structure, the silicon tetrahedron. These interpretations and the corresponding systematics may be called the first chapter of the crystal-chemistry of silicates. In the "second chapter," due to N. V. Belov, the principal part is played by the large cations Na, K, Ca and so forth, the coordination polyhedra (octahedra) of which have an edge longer than that of the silicon tetrahedron. A suitable length is that of the edge of the trigonal prism around the double tetrahedron of Si_2O_7, which thus plays the same part in the second chapter as the single tetrahedron SiO_4 did in the first. In particular, the Si_2O_7 groupings are decisive in the structure of natural and synthetic calcium hydrosilicates (cements). Like the pyroxene chains of $(Si_2O_6)_\infty$ or the double amphibole chains of $(Si_4O_{11})_\infty$ of the first chapter, we have as corresponding elements in the silicates of the second chapter the wollastonite chains of $(Si_3O_9)_\infty$ and xonotlite bands of $(Si_6O_{17})_\infty$. Where six-membered rings are characteristic of the chains, bands and nets of the first chapter, the analogous radicals of the second chapter are characterized by alternating four-membered and eight-membered rings. Even to-day, more than ten analogs in the second chapter correspond to two classical chains in the first chapter, and almost all of them have been discovered in the Soviet Union. Thus, the atomic structure of the basic components of the earth's crust, the silicates, are now presented to us in the form of a systematics which is not only more profound, but is also more complete and more universal.

In generalizing the work on the structure of silicates and other minerals, N. V. Belov has published in the past 12 years more than 80 "Outlines of Structural Mineralogy." In these "Outlines" and the recently published textbook "The Second Chapter of the Crystal Chemistry of Silicates," he is assiduously inculcating modern concepts on the atomic structure of minerals in wide circles of geochemists, mineralogists, and petrologists, and is providing these circles with fruitful ideas on the genetics of the structural relations between some minerals and others and the part played by various elements in these relations. His crystal-chemical generalizations on the nature of the properties of cement, structure of glasses, molecular screens and the geochemistry of the rare "heat-resisting" and light elements are not merely of theoretical value, but also of considerable practical value, witnessed by the constant contacts enjoyed by N. V. Belov with the workers of a number of branches of industry, particularly the glass and cement industries.

N. V. Belov's remarkable work on structural crystallography is embodied not only in his publications, but also in the museum of models, which he has formed at the Institute of Crystallography and which is one of the few of its kind in the world.

N. V. Belov is not only a prominent scientist, he is also an outstanding teacher and an untiring propagandist of his science. His activities as professor at the Moscow and Gor'kii universities have brought many young and talented physicists and mineralogists to the x-ray analysis, crystallography, electron analysis and other laboratories. His pupils are working in Moscow, Gor'kii, Baku, Novosibirsk, Kishinev, and other cities and towns of the Soviet Union. N. V. Belov is the editor of the Russian translations of a number of books by prominent foreign x-ray crystallographers. One must not overlook the editorial skill of N. V. Belov as a critic, through whose hands have passed scores of theses for the degree of doctor and candidate of science, the authors of which (and their supervisors) have received much valuable advice from the impartial criticism of the reviewer.

N. V. Belov's scientific results have been widely acclaimed not only at home but also abroad. It seems that among the modern generation of Soviet specialists in the field of structural analysis, it would be difficult to find any one who, in some shape or form, has not experienced the fruitful influence of N.V. Belov's ideas. This applies both to scientists currently in charge of the large laboratories of the many institutes of the country and to the capable coming generation.

The outstanding achievements of N. V. Belov were rewarded in 1946 by his election as corresponding member of the Academy of Sciences, USSR, and in 1953 as Academician. N. V. Belov is a laureate of the Stalin Prize, 1st Degree (1952). He is President of the National Committee of Soviet Crystallographers and Vice-President of the International Union of Crystallography.

In this momentous 70th birthday of our beloved Nikolai Vasil'evich Belov, which he celebrates at the prime of his creative powers, Soviet crystallographers wish him further success for the good of Soviet science.

Reprinted from Proceedings of the Academy of Sciences of the USSR (Doklady Akademii Nauk SSSR),
Geological Sciences Sections, Vol. 121, pp. 713-716, July-August, 1958

CRYSTAL STRUCTURE OF MICA-LIKE HYDROUS CALCIUM SILICATES:

OKENITE, NEKOITE, TRUSCOTTITE AND GYROLITE.

NEW SILICATE RADICAL $[Si_6O_{15}]_\infty$

Kh. S. Mamedov and Academician N. V. Belov

It is well known that the doubling of pyroxene chains $[SiO_3]_\infty$, gives amphibole bands $[Si_4O_{11}]_\infty$ consisting of hexagonal links which in their turn may polymerize into sheets $[Si_2O_5]_\infty$ with hexagonal links exclusively, such sheets being characteristic of a very large class of layered minerals — micas, clays and chlorites.

The pyroxenoid chains $[SiO_3]_\infty$ with different geometry than that of pyroxene chains (three Si-tetrahedra in a link instead of two), discovered by us in wollastonite [11], give double double chains with the formula $[Si_6O_{17}]_\infty$ and octagonal links (xonotlite bands [2]) which polymerize into sheets $[Si_2O_5]_\infty$ with a very different geometry.

Sheets with octagonal links alone are not possible according to elementary theorems of structural crystallography [3]. In the simplest xonotlite bands, octagonal and tetragonal links alternate, preserving tetragonal symmetry. The octagonal links make the network too open and are therefore compressed, but with preservation of tetragonal symmetry of the sheet as a whole. Such is the $[Si_2O_5]_\infty$ net in apophyllite, whose condensation along the normal to the sheet results in the pseudotetragonal three-dimensional frameworks $[(Si, Al)O_2]_\infty$ characteristic of feldspars [3].

In this deformation of the ideal tetragonal structure the parallelism of the condensed wollastonite and xonotlite chains is lost, but a number of hydrous Ca-silicates break down at high temperature into wollastonite fibers parallel to the b-axis of the sheet, which is exactly of the same length as the b-axis of wollastonite, and to account for this a different scheme of condensation of xonotlite bands has been proposed [4], according to which the bands are united in parallel position by sharing their side O-atoms. Such a net, as can be seen from Fig. 1-a, contains octagonal, hexagonal and rectangular links, which contradicts the principle of parsimony, and besides, the dominance of the mirror planes in the structure is unnatural since, generally speaking, these planes are not characteristic in crystallochemical patterns. The main objection to such a net, however, is the complete lack of agreement between its parameters and those established for natural and synthetic Ca-silicates of this type.

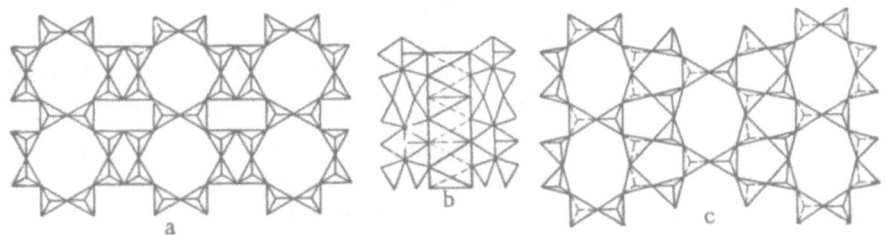

Fig. 1. Silicate radicals with wollastonite-xonotlite structure. a) Rhombic sheet
with symmetry pmm; b) union of wollastonite chain with a column of Ca-octahedra;
c) rhombic nets with symmetry cmm.

All difficulties are removed if it is considered that the basic units in the structure of wollastonite, xonotlite and the related minerals are endless columns of Ca-octahedra so arranged that the O—O edge of one octahedron is continued directly by the edge of the next one (Fig. 1-b). Such an edge is much longer than the edge of a Si-tetrahedron but smaller than the doubled altitude of the latter. The tetrahedra are inclined to the octahedral edges, and this makes it possible for two of them to be joined to it, while to the next octahedral edge corresponds the edge of one Si-tetrahedron plus the "excess" space created by the inclination of the preceding pair of tetrahedra. The edges of Ca-octahedra being alike, these two modes of union of Si-tetrahedra permit junction of adjacent xonotlite bands as shown in Fig. 1-c. The symmetry element in this structure is the glide plane so common in structural crystallography. The sheet consists of octagonal links alternating with twice as many pentagonal links known from the structures of melilite-gehlenite and epidote-zoisite groups.

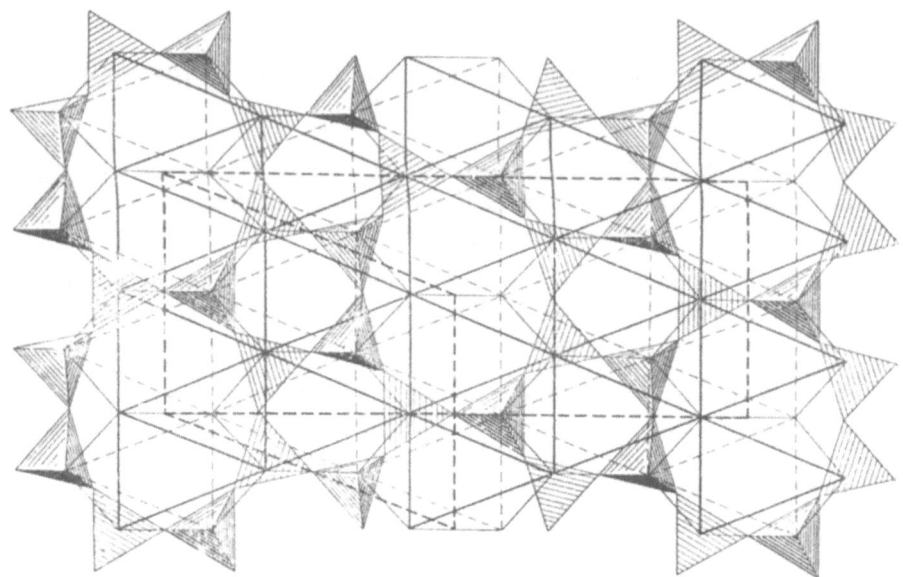

Fig. 2. Crystal structure of okenite: the silicate net $[Si_6O_{15}]_\infty$ above the closely packed layer of Ca-octahedra.

If the average O to O distance in the Si-tetrahedra is taken as 2.5 A, then in such a net a = 18.30 A, b = = 7.20 A for the orthogonal (centered) cell and a = 9.83 A, b = 7.20 A, γ = 111.5° for the oblique (primitive) cell. The primitive cell contains one xonotlite ring, i.e., 6 atoms of Si, and therefore the formula of the radical $[Si_2O_5]_\infty$ may be refined in the silicates under discussion to $[Si_6O_{15}]_\infty$.

By combining x-ray and electron diffraction methods, Gard and Taylor [5] obtained the following parameters for okenite, which represents the group of lamellar hydrous Ca-silicates: a = 18.30 A, b = 7.20 A, c = 21.33A, β = 105° for the (mica-like) centered cell, and a = 9.84 A, b = 7.20 A, c = 21.33A, α = 90°, β = 103.9°, γ = 111.5° for the primitive cell.

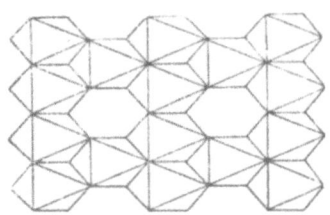

Fig. 3. Spinel pattern of stretched octahedra.

The primitive cell contains 9 formula units of $CaO \cdot 2SiO_2 \cdot 2H_2O$. The fibers are oriented parallel to the axis b = 7.20 A. The electron microscope clearly showed a perfect (001) cleavage. On being heated to 800°C, okenite loses water and breaks down into wollastonite whose fibers are oriented parallel to the b-axis of okenite (the b-axis of the two minerals being equal) and unoriented cristobalite.

$$CaO \cdot 2SiO_2 \cdot 2H_2O \rightarrow CaSiO_3 + SiO_2 + 2H_2O.$$

The formation of oriented wollastonite after dehydration is a good indication that the decomposing complex radical is a condensation of wollastonite chains; up to now, the best example of this structure was xonotlite [6].

Minerals	Gyrolite	Truscottite	Okenite	Nekoite
Symmetry and parameters of the unit cell	Pseudohexagonal $a = 9,72$ Å, $c = 6 \times 22,13$ Å	Pseudohexagonal $a = 9,72$ Å, $c = 18,71$ Å	Triclinic, $a = 9,84$ Å, $b = 7,20$ Å, $c = 21,33$, $\alpha = 90,0°$, $\beta = 103,9°$, $\gamma = 111,5°$	Triclinic, $a = 9,86$ Å, $b = 7,32$ Å, $c = 7,60$ Å, $\alpha = 103°54'$, $\beta = 86°12'$, $\gamma = 111°48'$
No. of formula units in the unit cell and their composition. Volume per one oxygen atom (V_O)	$24 [Ca_4Si_6O_{15}(OH)_2 \cdot 4H_2O]$ $22,7$ Å^3	$4 [Ca_3Si_6O_{15} \cdot 2H_2O]$ $22,2$ Å^3	$3 [Ca_3Si_6O_{15} \cdot 2H_2O \cdot 4H_2O]$ $22,2$ Å^3	$Ca_3Si_6O_{15} \cdot 2H_2O \cdot 4H_2O$ $23,5$ Å^3
No. of silicon atoms in the unit cell	144	24	18	6
No. of packets in the structure	18	3	3	1
No. of Si-atoms in a packet	8	8	6	6
Cleavage	(0001)	(0001)	(001)	(001)

T A B L E

It follows from the wollastonite and xonotlite structures [1, 2] that the thickness of a packet of Ca-octahedra — Si-tetrahedra in a single layer is 7.07 A; in contrast to micas and clays, the tetrahedra and octahedra point in opposite directions. Since in okenite c = 21.33 ≈ ≈ 3 × 7.07 A, its cell consists of three tiers of xonotlite packets and the content of the cell is $3[Ca_2Si_6O_{15} \cdot 6H_2O]$.

One ring in the xonotlite structure [2] contains 6 Ca-polyhedra in the cation layer. After condensation there remain only 4 polyhedra per ring. This corresponds to the difference in parameters 16.5 A in xonotlite and 18.30 A in okenite; Ca-octahedra in the latter are strongly stretched (Fig. 2).

If all four octahedra corresponding to one xonotlite ring are populated by Ca, then the CaO : SiO$_2$ ratio is 4 : 6, as in gyrolite ([7], see the Table). The ratio 1 : 2 = 3 : 6, characteristic of okenite ([5], Table), occurs if in the flat packing of octahedra only 3/4 of their number is occupied. The corresponding simplest pattern is the spinel pattern [8] with a similar centered orthogonal cell (Fig. 3).

To the closely packed layer of octahedra with 4 octahedra per unit cell correspond 8 atoms of O or its substitutes. It follows from the formula of okenite with 6 atoms of Si that only 6 of the 8 vertices of an octahedron belong to the Si-tetrahedra while the other two vertices must hold 2H$_2$O. The 6H$_2$O molecules in okenite are of two kinds — molecules of "constitutional" H$_2$O and 4 molecules of water which is probably zeolitic [9]. Geometrical analysis shows that the latter may occupy only those places which are indicated by broken circles in Fig. 4.

The final structural formula of okenite is $3 [Ca_2Si_6O_{15} \cdot 2H_2O \cdot 4H_2O]$.

The division of water into constitutional and zeolitic is confirmed by thermal analysis showing two sharp endothermic maxima.

Nekoite has the same formula but without the coefficient 3. This mineral with c = 7.60 A must be considered as consisting of single packets with like orientation. In gyrolite [6], on the contrary, most of the packets are so oriented as to make it pseudohexagonal.

The Ca-octahedra layers in gyrolite (as in uniaxial biotites) are continuous and this leads to the formula $Ca_4Si_6O_{15}(OH)_2 \cdot 4H_2O$ with the constitutional water being in the form of hydroxyl. Truscottite [10] $Ca_3Si_6O_{15} \cdot 2H_2O$, is also without zeolitic water, and this decreases its parameter c as compared with okenite. The two latter minerals become pseudohexagonal and the number of molecules in the packets increases by 33% as is shown in detail in the Table. It is remarkable that the changes in

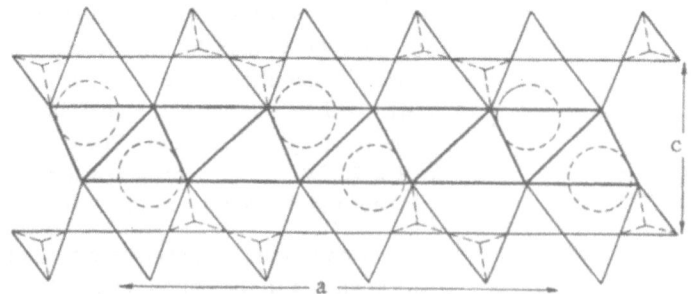

Fig. 4. Molecules of zeolitic H_2O in the okenite structure.

the basic dimensions of the structures related to the removal of zeolitic water, on the one hand, and tendency toward hexagonal symmetry on the other, are only slightly reflected in the value of V_O for each O atom, which for all hydrous Ca-silicates lies within the 22.2-23.5 A range (Table).

Institute of Chemistry, Academy of
Sciences, Azerbaidzhan SSR

Institute of Crystallography,
Academy of Sciences, USSR

Received April 29, 1958

LITERATURE CITED

[1] Kh. S. Mamedov and N. V. Belov, Doklady Akad. Nauk SSSR 104, No. 4 (1955).

[2] Kh. S. Mamedov and N. V. Belov, Doklady Akad. Nauk SSSR 107, No. 3 (1956).

[3] N. V. Belov, Kristallografiya 2, No. 3, 366 (1957).

[4] F. Liebau, Z. phys. Chem. 206, 73 (1957).

[5] J. A. Gard, and H. F. W. Taylor, Mineral. Mag. 31, 5 (1956).

[6] L. S. Dent and H. F. W. Taylor, Acta Crystal. 9, 1002 (1956).

[7] A. L. Mackay and H. F. W. Taylor, Mineral. Mag. 30, 80 (1953).

[8] N. V. Belov, Structure of Ionic Crystals, 1947.*

[9] J. D. Bernal, Usp. khim. 25, No. 5 (1956).

[10] A. L. Mackay and H. F. W. Taylor, Mineral. Mag. 30, 450 (1954).

---

*In Russian.

Reprinted from Proceedings of the Academy of Sciences of the USSR (Doklady Akademii Nauk SSSR),
Geological Sciences Sections, Vol. 121, pp. 725-727, July-August, 1958

THE CRYSTAL STRUCTURE OF FOSHAGITE — $Ca_8[Si_6O_{17}](OH)_6$

Kh. S. Mamedov and Academician N. V. Belov

The formula of this mineral, a member of a large class of fibrous, hydrated wollastonites [1, 2], was originally given [3] as $5CaO \cdot 3SiO_2 \cdot 3H_2O$; the synthetic product [4] is nearer to $3CaO \cdot 2SiO_2 \cdot H_2O$. A new analysis (analyst, Chalmers [5]) which took into consideration the presence in the sample of an admixture of SO_4-thaumasite and calcite gives the formula $4CaO \cdot 3SiO_2 \cdot H_2O$; the theoretical density for this formula corresponds well with the measured density of foshagite — 2.73.

Foshagite is a typical fibrous mineral and it is possible to obtain not only powder photographs, but also rotation photographs about the axis of a fiber. The repeat distance along the fiber axis is 7.35 A, and the half-distance of 3.68 A is also sharp, as in many other calcium silicates, hillebrandite, tobermorite and especially xonotlite [1, 2]. Foshagite resembles the latter in the high temperature of dehydration (650-750°C according to [5]). Powder photographs of foshagite heated to 600°C did not show any differences as compared with photographs of unheated specimens.

Like other calcium silicates mentioned above, foshagite heated to 800°C changes, after dehydration, into β-wollastonite whose fibers are oriented parallel to the fibers of the original mineral. The repeat distance of 7.35 A and the half-distance of 3.68 A are the same in both minerals (as well as in other minerals listed above).

By x-ray and electron diffraction methods Gard and Taylor [5] were able to determine the exact parameters of the monoclinic cell of foshagite: a = 10.32 A, b = 7.35 = 2 · 3.68 A; c = 14.07 = 2 · 7.03 A; β = 106°; they noted their resemblance to the parameters of β-wollastonite [6, 7]: a = 15.33 A; b = 7.28 = 2·3.64 A; c = 14.08 = = 2·7.04 A; β = 95°.

The electron microscope [5] revealed the presence of basal cleavage parallel to (001) and a less distinct cleavage parallel to (100). No special extinctions were observed, i.e., the mineral may belong either to group P2/m or to one of its subgroups P2 or Pm.

We have shown [7] that it is convenient in practice to describe β-wollastonite as having the cell with a = = 7.88 A, b = 7.28 A, c = 7.04 A, α = 90°, β = 95°16', γ = 103°22' and containing six units of $CaSiO_3$, i.e., $Ca_6[Si_3O_9]_2$. The corresponding cell of foshagite with a = 10.92 A, b = 7.35 A, c = 7.03 A, β = 106°4' contains $Ca_8[SiO_3]_6 \cdot (OH)_4$, i.e., there are two extra $Ca(OH)_2$ groups as compared with wollastonite.

Gard and Taylor, however, preferred a different interpretation. They based their argument on the fact that it is easy to separate in the wollastonite structure a quadrupled (quadrupled area as compared with the area of the primitive cell) cell which contains exactly three cells of foshagite (cf. Fig. 9 in [5]) and if the positions of Ca-ions in the cell are not disturbed, then a perfectly satisfactory coefficient of certainty is obtained for foshagite [8]. The foshagite cell, then, contains (6 × 4) : 3 = 8 Ca ions in four columns. As for the $[Si_3O_9]_\infty$ chains, there will not be 2 × 4 = 8 of them in the three cells, but only six, i.e., two for each foshagite cell, while the remaining two (as compared with wollastonite) fall out. After heating to 750°C these two chains $[Si_3O_9]_\infty$ return to every three foshagite cells and change them into four wollastonite cells. The possibility of this return is due to the fact that three adjacent foshagite cells disappear completely, two of the chains $[Si_3O_9]_\infty$ pass into the three remaining cells and four of them separate into individual (ortho) SiO_4-groups, with formation of unoriented small crystals of $β-Ca_2SiO_4$.

$$6Ca_8[Si_3O_9]_2(OH)_4 \xrightarrow{750°} 8Ca_3[Si_3O_9] + 12Ca_2SiO_4 + 12H_2O.$$

This scheme is too complicated, and it is unlikely that a quiet change from foshagite to wollastonite is possible if it is accompanied by the breaking up of three chains into individual tetrahedra and by the transfer of one-sixth of them into the neighboring cells. Gard and Taylor themselves remarked that their electron diffraction photographs did not indicate the formation of β-Ca_2SiO_4.

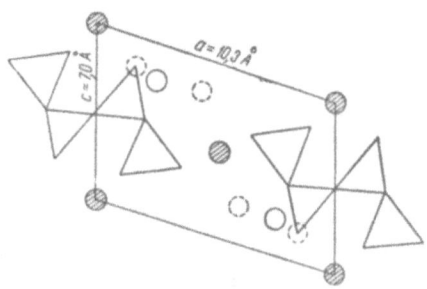

Fig. 1. Structure of foshagite. Plan.

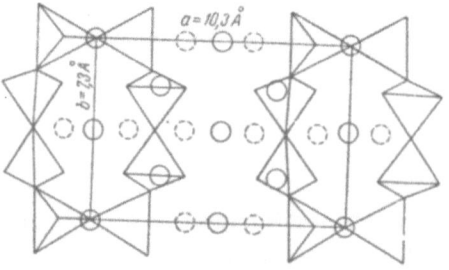

Fig. 2. Structure of foshagite. Front view.

The same authors emphasize the similarity in thermal behavior between foshagite and xonotlite which at about 700°C changes gradually into wollastonite by breaking up of the xonotlite double chains $[Si_6O_{17}]_\infty$ into single wollastonite chains.

Writing the formula of foshagite not as $4CaO \cdot 3SiO_2 \cdot H_2O$, but as $4CaO \cdot 3SiO_2 \cdot 1.5H_2O$, i.e., $8CaO \cdot 6SiO_2 \cdot 3H_2O$ (the first formula requires 4.3% of H_2O, the second 6.5%, but during the satisfactory chemical analysis cited above a considerable amount of H_2O was discarded as belonging to the water-rich thaumasite), we finally give this formula as $Ca_8[Si_6O_{17}](OH)_6$, indicating that foshagite contains the xonotlite radical — the double chain of Si-tetrahedra which was discovered by us in xonotlite in 1955 [1, 2]. There is only one such formula-unit in the cell. The thermal process between 650-750°C is analogous to that which occurs in xonotlite, i.e., it consists in escape of H_2O accompanied by the splitting of the double chain into single chains according to the equation

$$Ca_8[Si_6O_{17}](OH)_6 \xrightarrow{750°} 6CaSiO_3 + 2CaO + 3H_2O.$$

Another product of this reaction besides wollastonite is CaO, as was mentioned by Heller [9] in 1952.

Thus foshagite becomes a Ca-xonotlite

$$Ca_6[Si_6O_{17}](OH)_2 + 2Ca(OH)_2 = Ca_8[Si_6O_{17}](OH)_6.$$

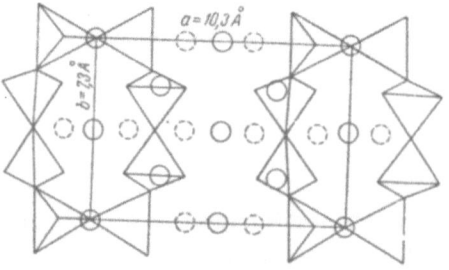

Fig. 3. Foshagite structure in Pauling's polyhedra.

In describing the structures of xonotlite and wollastonite (and also okenite, etc. [10]) we have pointed out more than once [11] that their basic pattern consists of infinite parallel columns of Ca-octahedra with associated chains of Si-tetrahedra and, as a rule, two Ca-octahedra and three Si-tetrahedra correspond to the length b = 7.3 A (one link). In the unit cell of β-$CaSiO_3$ there are three columns of Ca-octahedra and two $[Si_3O_9]_\infty$ chains; in xonotlite the same three columns of Ca-octahedra are associated with one double chain $[Si_6O_{17}]_\infty$ and one missing O is replaced stoichiometrically by two (OH) ions. In foshagite there are four columns of Ca-octahedra for each xonotlite chain (additional $2Ca(OH)_2$). The space for the additional column in the xonotlite structure is easily found in the center of the cell, as shown in articles [1, 2]. Figures 1 and 2 shows the plan and the front

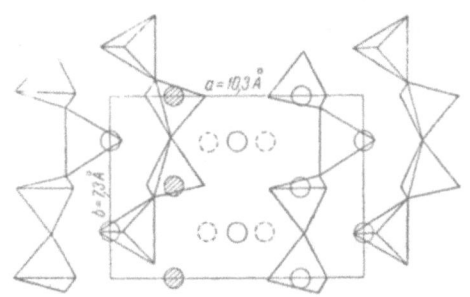

Fig. 4. Foshagite structure on the assumption of the existence in it of wollastonite chains $[Si_3O_9]_\infty$.

view of the foshagite structure; Fig. 3 shows the same plan in Pauling's polyhedra. Unlike xonotlite, in which two columns (out of three) are composed of Ca-prisms, in foshagite and in wollastonite all columns are built of Ca-octahedra. The (001) and (100) cleavages are clearly shown, for along these planes only the Ca-octahedra are separated. It should be noted that the distribution of the four columns of Ca-octahedra in this model corresponds to that assumed by Gard and Taylor on the basis of the wollastonite structure; i.e., here also the 25% coefficient of certainty, indicated by these authors, is approached.

It is not difficult (starting with the same four columns of Ca-octahedra) to build a model of foshagite directly (and completely) of wollastonite chains (Fig. 4), i.e., strictly according to the formula

$$2\,\{4CaO \cdot 3SiO_2 \cdot H_2O\} = Ca_8\,[Si_3O_9]_2\,(OH)_4,$$

but then the reaction at 750° is unexplainable, since the simple dehydration of Ca-minerals characteristic, for instance, of afwillite or tobermorite [5] requires temperatures not exceeding 500°C.

Institute of Chemistry
Academy of Sciences, Azerbaidzhan SSR

Institute of Crystallography
Academy of Science, USSR

Received May 7, 1958

LITERATURE CITED

[1] Kh. S. Mamedov and N. V. Belov, Doklady Akad. Nauk SSSR 104, No. 4 (1955).

[2] Kh. S. Mamedov and N. V. Belov, Zap. Vsesoiuzn. mineral. obshch. 85, No. 1 (1956).

[3] A. S. Eakle, Am. Mineral. 10, 97 (1925).

[4] W. Jander and B. Franke, Z. anorg. Chem. 247, 161 (1941).

[5] J. A. Gard and H. F. W. Taylor, Am. Mineral. 43, 1 (1958).

[6] J. W. Jeffery, Acta Cryst. 6, 321 (1953).

[7] Kh. S. Mamedov and N. V. Belov, Doklady Akad. Nauk SSSR 107, No. 3 (1956).

[8] W. Taylor, Paper Read at the Berlin Symposium on Crystal Chemistry, March 1958 [Russian translation].

[9] L. Heller, Proc. 3-d Int. Symp. Chemistry of Cement, 1952, p. 237.

[10] Kh. S. Mamedov and N. V. Belov, Doklady Akad. Nauk SSSR 121, No. 4 (1958).*

[11] N. V. Belov, Zap. Vsesoiuzn. mineral. obshch. 86, No. 2 (1957).

---

*Original Russian pagination. See C. B. Translation.

Reprinted from Proceedings of the Academy of Sciences of the USSR (Doklady Akademii Nauk SSSR),
Geological Sciences Sections, Vol. 123, pp. 1035-37, November-December, 1958

THE CRYSTALLINE STRUCTURE OF TOBERMORITES

Kh. S. Mamedov and Academician N. V. Belov

Tobermorites are fibrous and, at the same time, tabular, mica-like Ca hydrosilicates (with perfect prismatic cleavage), which are similar to montmorillonite and vermiculite in being able to take on large quantities of water while remaining (mono) crystalline and with a corresponding increase in but one of the parameters of the unit cell. The essential difference between montmorillonite and vermiculite is the intermittent, sharply step-like increase in the amount of H_2O, which determines the different terms used for the various stages.

All tobermorites are characterized by orthorhombic symmetry [1]. The parameters $\underline{a}$ and $\underline{b}$ in known natural and easily synthesized varieties (silicate bricks) are identical: a= $2 \cdot 5.65 = 11.3$ A, b = $2 \cdot 3.65 = 7.30$ A. The parameter $\underline{c}$, corresponding to the plane of perfect cleavage (001), is 28 A for 14-A hydrate (plombierite), 22.6 A for 11-A hydrate (tobermorite proper), 20.5 A for 10 A, and 18 A for 9 A (riversideite). The Angstrom unit standing before the "hydrate" describes the characteristic (interplanar distances) d_{002}, which distinguish tobermorites (by x-ray powder photographs; other values of $\underline{d}$ are almost identical), and the doubled value gives the period of recurrence (c) in the direction normal to the alternating layers.

All the enumerated hydrates are converted to 9-A hydrate on heating at 250-300°. On treating the 14-A hydrate with water, $\underline{d}$ increases to 29.2A. The 9-A hydrate is unstable and in air gradually changes to the 11-A hydrate. The 9-A hydrate is stable up to 700°, but above 800° it is shown to convert to fibers of wollastonite, with parallel orientation, having the same b = $2 \cdot 3.65 = 7.30$ A.

If we do not consider the quantity of interlayer water, the compositions of all the tobermorites are identical: the orthorhombic cell contains four formal units consisting of $5CaO \cdot 6SiO_2 \cdot xH_2O$. The ratio $SiO_2 : H_2O$ is about 0.5 for 9-A hydrate, about 1 for 11-A hydrate, and 2-2.5 for 14-A hydrate [1].

In contrast to the others, this is shown to be "tobermorite from Loch Eynort" [2]. Formally all its parameters correspond to 11-A hydrate (see below on the parameter $\underline{a}$): it also converts to oriented fibers of wollastonite, but the principal cleavage is not (001) but (100), and samples of it do not change to 14-A hydrate even after being crushed to powder and treated with H_2O for many hours.

Megaw and Kelsey [3], from studies of single crystals measuring tens of millimeters, have described the structure of the best known tobermorite, the 11-A hydrate, which agrees satisfactorily with the diffraction pattern. They began with the composition $5CaO \cdot 6SiO_2 \cdot 5H_2O$, indicated by McConnell [4], and with the concept of the basic role of the wollastonite chains of $[SiO_3]_\infty$, which are characteristic of the material thermally transformed from it. In regard to the Ca silicate afwillite [5], occurring a short time earlier than the structure above determined, it was proposed that part of the O in the $[SiO_3]_\infty$ chains is replaced directly by the hydroxyl groups OH. The chains determine the fibrous nature of tobermorite; the tabular form of the mineral is due to condensation of the wollastonite chain into layers, similar to the condensation of the amphibole chains into a talc net. These latter, as is well known, are single-layer planar formations of Si tetrahedra, such as the nets (with partial replacement of Si by Al tetrahedra) in micas and chlorites (clays) and also (with a somewhat different lattice) such Ca silicates as okenite and gyrolite [6]. The silicon-oxygen nets in tobermorites, according to Megaw and Kelsey, are like a box crimped from both sides. In the 9-A hydrate, the junction of one of the two nets occurring at the period $\underline{c}$ emerges in the space between the seams of the neighboring layers and there is always sufficient space between the layers for Ca cations. An ever increasing number of H_2O molecules appear between the layers of the more thoroughly hydrated tobermorites.

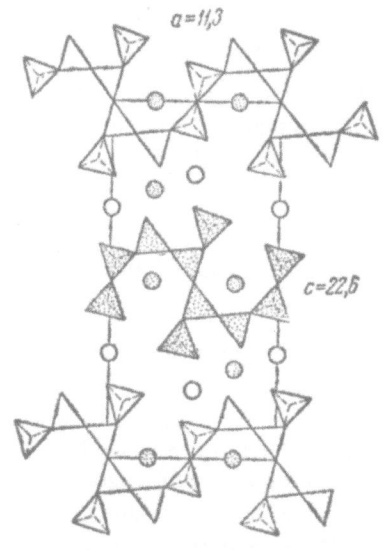

Fig. 1. Tobermorite. Projection viewed along the xonotlite band, coupled in crimped silicon-oxygen nets.

From this scheme of Megaw and Kelsey we go on to the actual structure of the layers of tobermorite, taking into account the experience of our last interpretation, which showed (foshagite [7], okenite [6], etc.) that if the decomposition product of the Ca hydrosilicates at about 750° is fibrous wollastonite with solitary chains of $[Si_3O_9]_\infty = 3SiO_3$, then double silicon-oxygen chains of xonotlitic $[Si_6O_{17}]_\infty$ [8] (in a hydrothermal synthesis of Ca silicates) are produced. In xonotlite itself these chains are discrete and are arranged in two's with a recurring spacing (in the plane of the band) of a = 16.5 A. In okenite the double chains are discrete and they intergrow with each other in a parallel position, but with a displacement of the half period $\underline{b}$, i.e., along the axis of the chains. The okenite net, as in talc-micas, is perfectly planar. In tobermorite, the xonotlite bands, being condensed, are adjusted to each other with the formation of seam-crimps, about which we spoke above. Details of this condensation are shown in Fig. 1, which gives the projection of the tobermorite structure viewed along the chain (end views of the chains), and in Fig. 2, which gives a projection with adjusted rings viewed from the normal to the plane of the net.* This shows rather graphically that, by mutual adjustment of the chains in tobermorite, the period a = 11.22 A is but 2/3 the corresponding period in xonotlite.

It is not difficult to believe that during the condensation corresponding to Figs. 1 and 2, each link of the xonotlite chain is joined by three atoms of O (two during condensation merely of the closing ring and one by the adjustment), i.e., the formula of the developing net will be

$$2\,[Si_6O_{17}]_\infty^{10-} - 3O^{2-} = [Si_{12}O_{31}]_{\infty,\infty}^{14-},$$

and the chemical formula for the constant part of tobermorite becomes instead

$$4\,\{5CaO \cdot 6SiO_2 \cdot x\,H_2O\} \rightarrow 2\,\{Ca_{10}\,[Si_{12}O_{31}]\,(OH)_6\}.$$

For the 9-A hydrate with the decreased quantity of water and the ratio (Ca)Si : $H_2O \approx 0.5$, beginning with the model, we should write

$$2\,\{Ca_{10}\,[Si_{12}O_{31}]\,(OH)_6 \cdot 3H_2O\}.$$

For tobermorites richer in water:

(11 Å) $Ca_{10}\,[Si_{12}O_{31}]\,(OH)_6 \cdot 8H_2O$,
(14 Å) $Ca_{10}\,[Si_{12}O_{31}]\,(CH)_6 \cdot 18H_2O$.

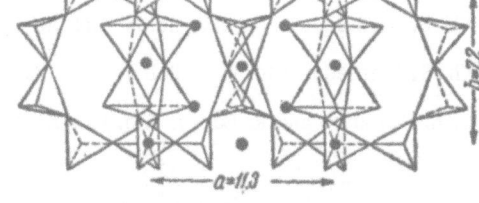

Fig. 2. Tobermorite. The mechanism of chaining and the mutual position of the xonotlite rings.

Figure 1 shows the approximate arrangement in the cell of the 11-A hydrate appearing in each layer of 12 Si tetrahedra (4 overlapping pairs) and also of 20 cations (10 overlapping pairs) of Ca. Twelve of the hydroxyl

*G. L. Kalousek and R. Roy [9], studying tobermorite and xonotlite in the infrared region, discovered a striking similarity in the absorption band in the region of 8-15 μ. However, the band at 6.2 μ is characteristic of tobermorite and is generally associated with interlayer H_2O, which is absent in xonotlite. The band at 2.9 μ in tobermorite corresponds to that in xonotlite at 2.75 μ, indicating the different nature of the OH bond of the hydroxyl group.

groups OH (in the entire cell) are all found under the nonoverlapping atoms of O in the horizontal junctions of the xonotlite rings. It is more difficult to indicate the position of the 16 (for the cell) units of interlayer H_2O.

The paradox with the Loch Eynort tobermorite is satisfactorily resolved when we turn our attention to the fact that the electron diffraction pattern in [2] for this tobermorite clearly shows that its period a is not equal to $2 \cdot 5.5 = 11$ A, as in ordinary tobermorites, but to $3 \cdot 5.5 = 16.5$ A. Thus the parameters a and b of the Loch Eynort tobermorite are shown to be a repetition of the parameters for xonotlite in which the parameter c is three times as large.

The concept thus arising that the Loch Eynort tobermorite is fundamentally similar to xonotlite finds immediate support in other distinctive features in the x-ray prints of the first mineral.

In [8] there is a detailed discussion to show that in both Ca silicates with precisely established chain structure, wollastonite and xonotlite, there is a common characteristic displacement of neighboring silicon-oxygen chains to the half period of b (along the axis of the chains) in a direction opposite to the normal displacement in the monoclinic cell. "Errors" of this kind produce streaks on the x-ray photograph at right angles to the b • axis, which indicates that the crystal has strict periodicity in but one direction. Such streaks are also characteristic of the prints from the Loch Eynort tobermorite, i.e., its structural units are endless in only one direction. Having established the xonotlite chains in the structure of the Loch Eynort tobermorite, we may represent the general formula for tobermorite in the form

$$2 \{Ca_5 [Si_6O_{17}] \cdot 5H_2O\} = Ca_4Ca_6 [Si_6O_{17}]_2 \cdot 10H_2O.$$

Thus, the Loch Eynort tobermorite is a three-layered xonotlite, having lost one $Ca(OH)_2$ but with an excessive 5 H_2O, and is an intermediate member between ordinary tobermorite and xonotlite; at least its characteristic features do not contradict this view.

LITERATURE CITED

[1] L. Heller and H. F. W. Taylor, Crystallographic Data for the Calcium Silicates, London, 1956.

[2] J. A. Gard and H. F. W. Taylor, Mineral. Magaz. 31, 361 (1957).

[3] H. D. Megaw and C. H. Kelsey, Nature 177, 390 (1956).

[4] J. D. C. McConnell, Mineral. Magaz. 30, 293 (1954).

[5] H. D. Megaw, Acta Crystall. 5, 477 (1952).

[6] Kh. S. Mamedov and N. V. Belov, Doklady Akad. Nauk SSSR 121, No. 4 (1958).*

[7] Kh. S. Mamedov and N. V. Belov, Doklady Akad. Nauk SSSR 121, No. 5 (1958).*

[8] Kh. S. Mamedov and N. V. Belov, Zap. Vsesoiuzn. mineral. obshch. pt. 85, No. 1 (1956).

[9] G. L. Kalousek and R. Roy, J. Am. Ceram. Soc. 40, 236 (1957).

Institute of Chemistry, Academy of Sciences,
Azerb. SSR
Institute of Crystallography, Academy of Sciences,
USSR

Received August 1, 1958

* See C. B. Translation.

Reprinted from Soviet Physics—Doklady, Vol. 4, pp. 20-23, August, 1959

THE CRYSTAL STRUCTURE OF LAWSONITE

I. M. Rumanova and T. I. Skipetrova

Institute for Crystallography, Academy of Sciences of the USSR

(Presented by Academician N. V. Belov, July 24, 1958)

The structure of lawsonite $CaAl_2[Si_2O_7](OH)_2 \cdot H_2O$ was determined as part of a program of investigations on silicate structures which is being carried out in the x-ray laboratory of the Institute for Crystallography of the Academy of Sciences of the USSR.

In the first x-ray examination of lawsonite (Gossner and Mussgnug [1]) the parameters of the orthorhombic cell (a = 8.85 A, b = 5.87 A, c = 13.22 A) and the space group of the crystal (D_{2h}^{17} = Ccmm) were determined. Wickman [2], in 1947, carried out a more detailed analysis; his parameters (a = 8.88 A, b = 5.75 A, c = 13.30 A) were in agreement with the previous ones, but he attributed the noncentrosymmetric space group D_2^5 = C 222_1 to lawsonite, although the model of the structure proposed in [2] was clearly centrosymmetric. If we consider the factual material in [2], in particular the table of intensities, it is readily seen that with odd l, not only F_{00l} = 0, but also F_{02l} = 0, i.e., there is very likely a c glide plane parallel to (100).

The structure analysis in [2] was carried out by the method of trial and error from moving films of the zero, first, and second layer lines obtained by rotation about b, taken using Cu K_α radiation; the intensities were estimated visually using a five-point scale. In the structure, chains of Al octahedra are linked together by diorthosilicate groups $[Si_2O_7]$. The Ca cations lie in the spaces, with a five-point figure as coordination polyhedron (a regular pyramid with square base), which is always improbable. The facts mentioned and the extremely approximate estimation of the intensities made a fresh determination of the structure of lawsonite desirable.

In our analysis, x-ray rotation photographs about a, b, and c were taken using MoK$_\alpha$ radiation, and also moving films of the layer lines obtained by rotation about b (0, 1, 2) and a (0); the intensities of the reflections were estimated visually by the degree of blackening. The cell parameters found correspond well with the earlier results [1, 2]: a = 8.83 A, b = 5.80 A, c = 13.20 A. The systematic absences observed (F_{hkl} = 0 if h + k = 2n + 1, F_{0kl} = 0 if l = 2n + 1) mean that the diffraction group is Cc − −; three orthorhombic space groups are therefore possible: Ccm2, Cc2m, or Ccmm. Intensity statistics [3], applied to the h1l reflections were used to solve the problem of the centro- or noncentrosymmetric nature of the cell. Averaging of the structure amplitudes and their squares was used for all reflections out to sin $\vartheta/\lambda \leq 0.8$ A^{-1} (of these there were 114 nonzero F_{h1l} and 30 zero). If the 0 intensities were added to these last, we obtained $\overline{F}_{h1l}^2 / \overline{F_{h1l}^2}$ = 0.575; if, instead of the zero values, half the minimum value of F_{h1l} as a function of sin ϑ/λ is taken, then $\overline{F}_{h1l}^2 / \overline{F_{h1l}^2}$ = 0.635, which corresponds to a centrosymmetric electron density distribution in the crystal. Experiments to detect a piezo-effect were negative in their results, i.e., were also in favor of a center of symmetry, so that we consider the space group of lawsonite to be Ccmm = D_{2h}^{17}, which, as has been mentioned, was previously proposed by Gossner and Mussgnug.

In the x-ray rotation photographs around all three axes a, b, and c, the intensities of the odd layer lines were noticeably weaker than those of the even ones. This can be explained in terms of the space group Ccmm quite naturally, if some of the cations lie in the 8 centers of symmetry on the 2_1 axes in the eight-fold (d) positions, which contribute only to structure factors of reflections with k, l = 2n (h = 2n). Such cations can only be the Al, of which there are eight in the cell (it is impossible to site the silicon atoms, which are always surrounded by a tetrahedron of 4 O atoms, at a center of symmetry, and there are only 4 Ca cations in the cell). The 4 Ca evidently occupy the four-fold (c) positions at the intersection of the two mirror planes of symmetry, since the

TABLE 1

Atom and No. in cell	Position and point symmetry	x/a	y/b	z/c	Atom and No. in cell	Position and point symmetry	x/a	y/b	z/c
4 Ca	(c) mm	0.338	0	0.25	16 O_{II}	(h) 1	0.379	0.263	0.118
8 Al	(d) $\bar{1}$	0.25	0.25	0	8 O_{III}	(f) m	0.138	0	0.060
8 Si	(f) m	−0.020	0	0.134	8 OH	(f) m	0.640	0	0.058
4 O_I	(c) mm	0.051	0	0.25	4 H_2O	(c) mm	0.609	0	0.25

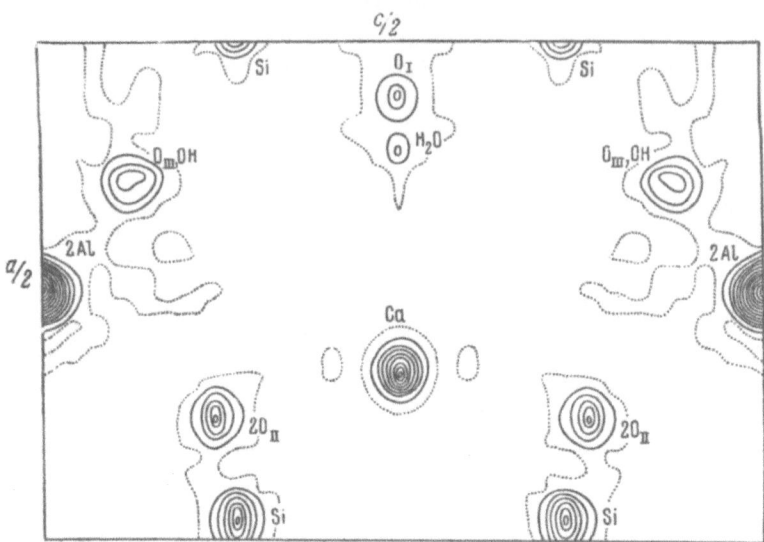

Fig. 1. Projection of the electron density on the xz plane. The dotted line represents a value $F_{000}/S_{xz} = 5.39$ e/A^2.

intensities of reflections on the layer lines with $l = 2n + 1$ are no weaker than on those with $h = 2n + 1$ and $k = 2n + 1$, whereas for the other four-fold positions (a) and (b) in two groups of four, of centers of symmetry, on the two-fold axes 2) $F_{hkl} = 0$ with $l = 2n + 1$.

With the Al fixed at the centers of symmetry, and especially since they appear with double height in the xz projection, the coordinates of the Ca and Si cations were found without difficulty from the Patterson projections

$$P(u, w) \text{ and } C_2(u, w) = \int_0^b P(u, v, w)\cos 4\pi v \, dv = \frac{1}{S_{uw}} \sum_{h,l=-\infty}^{\infty} F_{h2l}^2 \cos 2\pi (hu + lw), \text{ in which the interatomic}$$

vectors of greatest weight are (2Al−Ca), (2Al−Si) and (Ca−Si). The positions of Ca in (c) and Si in (f), found from P(u, w) and C_2(u, w), correspond, to a first approximation, to the results in [2].

The final analysis of the structure was then carried out by using the statistical method of Zachariasen [4]. The signs of 172 nonzero amplitudes F_{h0l} were calculated from the approximately determined coordinates of the Al, Ca and Si. 60 of them, with the largest unitary amplitudes $|U_{h0l}|$, for which it was possible to assume that the sign of F_{h0l} calculated from the cations does not change when the O atoms are taken into consideration, formed the basic set of reflections [5, 6]. The signs of the remainder were found from the equation $S_H = \overline{= S(S_{K_i} \cdot S_{H+K_i})}$. Satisfactory results were obtained for 136 S_{h0l} with 36 S_{h0l} undetermined. Having 136 F_{h0l} with signs, we constructed the electron density projection $\sigma(x, z)$, in which all the O atoms appeared, and from which the positions of the Ca and Si were refined. Moreover, using the coordinates of all the atoms, the signs of the 172 nonzero F_{h0l} were recalculated for the construction of a refined $\sigma(x, z)$ synthesis, on the electronic calculating machine "Strela," which is shown as a contour map in Fig. 1. There is practically no false detail in it. Judging from the height and shape of the peak, the atoms O_{III} and the OH group do not lie exactly on top of one

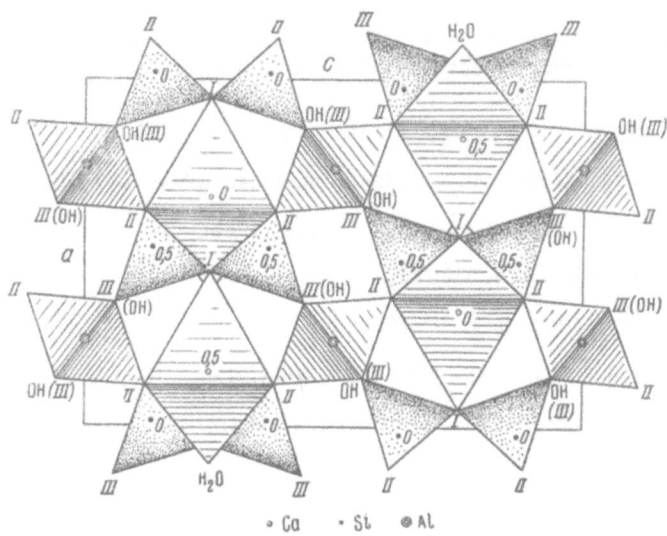

° Ca   · Si   ⊚ Al

Fig. 2. Projection of the structure of lawsonite on the xz plane
in coordination polyhedra.

another, but diverge slighly in the xz projection. With the exception of Al and O_{II}, all the other atoms in the projection are resolved, and consequently occupy special positions on the mirror planes y = 0 (y = $\frac{1}{2}$), which is confirmed by the C_2 (u, w) synthesis.

The coordinates of the atoms in the asymmetric unit of lawsonite, given in Table 1, were taken from this refined projection σ (x, z) (the y coordinate of O_{II}, which together with O_{II}^1 forms the vertical edge of the Si tetrahedron, was derived from crystallochemical considerations). If the coordinates obtained are compared with the results of [2], where the space group was D_{2h}^{17} = Ccmm, it is easy to see their similarity, but also the noticeable inaccuracy of the latter. The greatest difference is in the position of the water groups: the displacement along the a axis exceeds 0.6 A. For the remaining atoms the divergences are 0.21 A (O_I), 0.15 A (O_{II}), 0.13 A (Ca), etc. The reliability index, calculated from the 172 nonzero F_{h0l}, R_{h0l} = 16.7% (the reliability index calculated from the coordinates in [2], R_{h0l}^W = 33.4%).

In Fig. 2 is shown a projection of the structure of lawsonite on the xz plane in coordination polyhedra. The basic architectural motif is made up of infinite columns of Al octahedra extending along the two-fold screw axes parallel to b. Each Al lies at the center of an oxygen octahedron. There are two of these octahedra to the height of the cell, and they are joined by a common horizontal edge (O_{III} – OH) lying in the mirror plane. In each cell there are four columns of Al octahedra which are joined to each other by the diorthosilicate groups [Si_2O_7]. The Si–O distances lie in the narrow range 1.65-1.69 A (against 1.57-1.71 A in [2]); the O–O edges of the Si tetrahedron are within the limits 2.60-2.78 A (2.60-2.94 A in [2]), the Al–O distances are within the limits 1.92-1.95 A (1.92-1.98 A in [2]). The valence angle Si–O–Si of the diortho-groups is 135° 24' (148° 52' in [2]).

The spaces between the Al octahedra and the diortho-groups are occupied by the large Ca cations, around which are formed slightly distorted O (H_2O) octahedra which form the link between the neighboring Al columns. Six Ca–O (H_2O) distances correspond well to the sum of the ionic radii (2.4 A) and are within the very narrow limits for a "loose" Ca cation (2.36-2.53 A). The two next largest distances, Ca–O_{III} = 3.06 A, greatly exceed the preceding ones and therefore O_{III} cannot be included in the Ca polyhedron. Five-fold coordination is given in [2] for the Ca, since 5 Ca–O distances are within the limits 2.4-2.62 A and two further Ca–O distances are greater than 3 A (3.04 A), and in exactly the same way Ca–H_2O = 3.07 A.

The structure explains the cleavage in lawsonite well: this is perfect along (001) and less perfect along (100). The (001) cleavage plane passes through the Al columns; with cleavage along (100) the bonds in the Ca octahedron are broken in addition to the bonds in the Al columns, which means that the cleavage along (100) is slightly poorer than along (001).

The positive optical sign of lawsonite corresponds with columns of Al octahedra extended parallel to b.

We are deeply grateful to Academician N. V. Belov for valuable comments made in assessing this work, and also to R. G. Matveeva for providing moving-film photographs of layer lines of rotation about b.

LITERATURE CITED

[1] B. Gossner and F. Mussgnug, Zbl. Mineral. Geol. und Paläontol. A, 419 (1931).

[2] F. E. Wickman, Ark. Kemi Mineral. Geol. 25, A, 2, 1 (1952).

[3] A. J. C. Wilson, Acta Cryst. 2, 5, 318 (1949).

[4] W. H. Zachariasen, Acta Cryst. 5, 1, 68 (1952).

[5] I. M. Rumanova, Doklady Akad. Nauk SSSR 98, 3, 399 (1954).

[6] E. G. Fesenko, I. M. Rumanova and N. V. Belov, Kristallografiya 1, 2, 173 (1956).

Received July 21, 1958

Reprinted from Soviet Physics—Crystallography, Vol. 4, pp. 146-157, February, 1960

THE DETERMINATION OF THE STRUCTURE OF SEIDOZERITE

V. I. Simonov and N. V. Belov

Institute of Crystallography, Acad. Sci. USSR

The crystal structure of the $Zr-Ti$ silicate seidozerite has been elucidated by superposition on the (x, z) Patterson projection and by statistical determination of the signs of the F_{0kl} reflections. From the structure the expanded chemical formula of seidozerite is given as $Na_4MnTi(Zr_{1.5}Ti_{0.5})O_2(F, OH)_2[Si_2O_7]_2$ and its diorthosilicate nature (containing the group Si_2O_7) is revealed, despite the orthosilicate stoichiometric formula.

I. The new zirconium — titanium silicate seidozerite was discovered by E. I. Semenov in the Lovozerskii massif (Kol'skii P-ov) and was named after the place where it was first discovered, the Seidozero region [1]. In accordance with an analysis carried out by M. E. Kazakova, E. I. Semenov wrote the chemical formula of seidozerite in the form $Na_2(Mn_{0.5}Ti_{0.75}Zr_{0.75})Si_2O_8(F, OH)$. The monoclinic symmetry of the mineral, which was fairly obvious from optical goniometry [1], was confirmed, first from Laue photographs and later from an analysis of x ray rotation photographs and moving films of the zero, first and second layer lines (Mo radiation) about the axes [100] and [010]. The systematic absences among the $h0l$ reflections, with $l = 2n + 1$, observed in the films, are of the Laue class $C_{2h} = 2/m$ (the unique axis being $\underline{b}$ and the primitive ac parallelogram in the unique plane) and lead to the x ray group $2/mP-/c$ with a $\underline{c}$ glide plane, i.e., to two possible space groups $C_s^2 = Pc$ and $C_{2h}^4 = P2/c$.

The parameters of the unit cell of seidozerite are: $a = 5.53 \pm 0.2$ A, $b = 7.10 \pm 0.03$ A, $c = 18.30 \pm 0.10$ A and $\beta = 102°43' \pm 1'$. The angle β was measured on an optical goniometer. This setting corresponds to the three pinacoids characteristically appearing in the habit of the crystals [1]. If we select as axes the two shortest translations in the plane of symmetry, we arrive at the setting $2/mP-/n$ with the parameters: $a_1 = 5.53$, $b_1 = 7.10$, $c_1 = 17.90$ A and $\beta_1 = 85°19'$. The new axes are related to the old by the simple relationship $\begin{pmatrix} 1 & 0 & 0 \\ 0 & 1 & 0 \\ 1 & 0 & 1 \end{pmatrix}$.

In the subsequent work we used the first setting, which proved to be the most natural one in the analysis.

The conclusion that seidozerite was probably centrosymmetric, based on the crystal habit [1] and on the absence of a piezo effect (the experiments were kindly carried out by V. A. Koptsik), was confirmed by a statistical analysis of the intensities I_{h0l} by the method described in [2]. Another indication of centrosymmetry in seidozerite was obtained from the appearance in the Patterson projections of the characteristic trios of peaks described in [3]. We were therefore able to assign seidozerite, with reasonable certainty, to the holohedral group $P2/c = C_{2h}^4$.

From the density of seidozerite $d = 3.47$ (according to Semenov [1]) and from the volume of the unit cell, it appears that the latter contains 4 (3.87) formula units of $Na_2(Mn_{0.5}Ti_{0.75}Zr_{0.75})Si_2O_8(F, OH)$.

Thus, in the unit cell of seidozerite there are 3 Ti atoms and 3 Zr atoms in the space group $P2/c$, which contains only two-fold and four-fold positions (in the group Pc only two-fold), which leads to the interesting necessity, from a crystal- and geochemical standpoint, of filling certain crystallographically equivalent positions

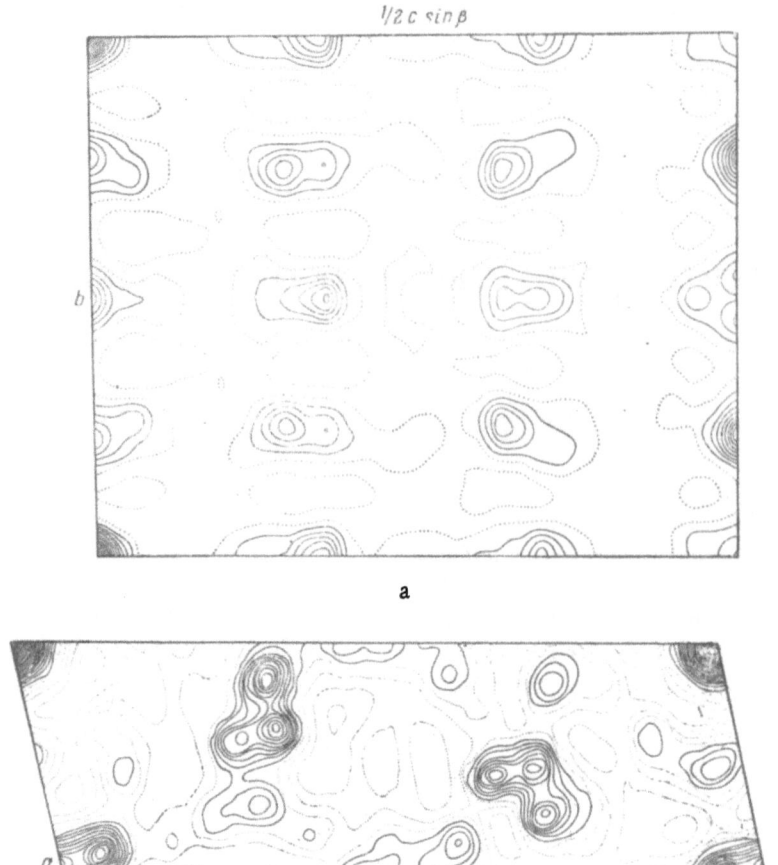

Fig. 1. Patterson projections for seidozerite: a) $p(y, z)$; b) $p(x, z)$.

by both Zr and Ti, whose atomic radii differ by $\sim 30\%$ ($r_{Ti}^{+4} = 0.64$, $r_{Zr}^{+4} = 0.82$ A). * Ti does not appear in the Fersman "star" of isomorphs for Zr [4], but it does occur persistently as such in all works on the Lovozerskii massif. Thus, the richest Zr lovenites may contain up to 30% of ZrO_2, and a significant part of this, as much as half, is commonly replaced by TiO_2 [5].

E. I. Semenov attributed seidozerite to the velerite group which contains the following silicates of fairly complex structure [6]: velerite $(Ca, Na)_3 (Zr, Nb) [SiO_4]_2F$, rosenbuschite $(Na, Ca)_3$, $(Fe, Ti, Zr) [SiO_4]_2F$, guarinite $Ca_2NaZr[SiO_4]_2F$ and lovenite $(Na, Ca, Mn\cdot\cdot)_3Zr[SiO_4]_2F$. No crystal structure has as yet been found for any of the minerals listed. This fact, together with the large number of minerals of such elements as Zr, Ti and Nb made the structure analysis interesting, not only from the geo-(crystal-)chemical point of view, but also from a practical standpoint.

In the analysis of the structure of seidozerite, 378 nonzero $h0l$ and 331 $0kl$ reflections were used. The intensities were assessed visually and converted to $|F_{hkl}|$ in the usual way. Owing to the large size of the $\underline{c}$

*For r_{Zr}^{+4} we took the figure of Bokii and Belov of 0.82 A. If the Goldschmidt value of $r_{Zr}^{+4} = 0.87$ is taken then the difference is as large as 36%.

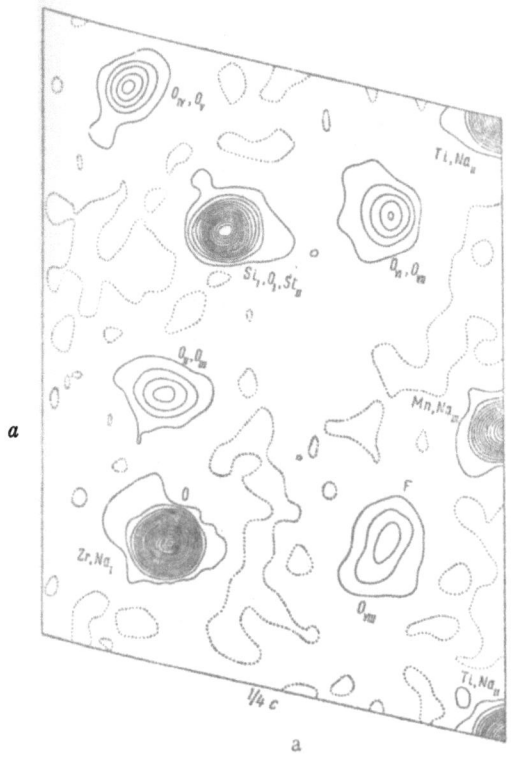

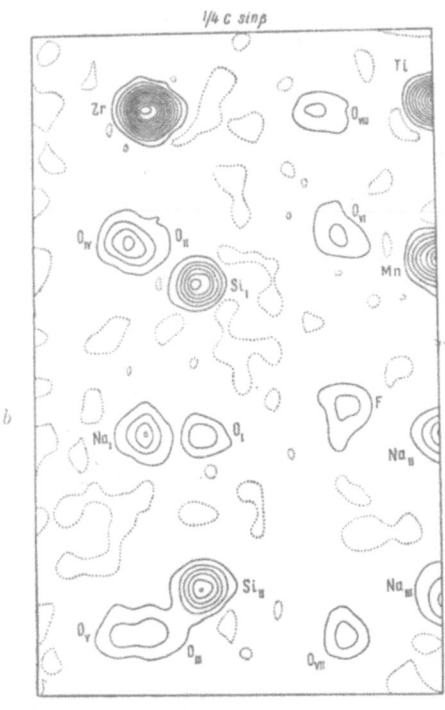

Fig. 2. Final electron density projections for seidozerite. a) $\rho(x, z)$; b) $\rho(y, z)$.

axis, 15 reflections of low $\sin\theta/\lambda$ were obscured on the $h0l$ and $0kl$ moving films which were taken with Mo radiation, and their intensities were assessed from x ray photographs taken with Cu radiation. The linear absorption coefficient $\mu = V^{-1}\Sigma n_i\mu_i^a(\mu_i^a$ is the atomic absorption coefficient for $\lambda = 0.71$ A [7]), calculated for seidozerite, is 35 cm^{-1}. The $h0l$ zero layer line moving film was taken from a crystal with a roughly rectangular cross section 0.15×0.20 mm. The $0kl$ reflections were measured from scattering from a crystal with cross section 0.3×0.5 mm. No correction for absorption was applied in the structure determination. This introduces an error of up to 10% in the $|F_{h0l}|_{obs}$ and up to 17% in $|F_{0kl}|_{obs}$.

II. The Patterson projections $P(y, z)$ and $P(x, z)$ for seidozerite are shown in Fig. 1. It was hoped, owing to the presence of the heavy Zr (Z = 40), to find the Zr—Zr peaks and to carry out superpositions from them [8-13]. It was natural to begin with the Patterson projection down the shortest translation $p(y, z)$. The peaks on it are arranged at the points of an almost regular rectangular net $b/4 \times c/6$, and differ in height by a factor of not more than two. Because of this regularity, minimalization from any peak preserved, with changed heights, all the peaks in $p(y, z)$ and it was impossible to draw convincing conclusions as to the structure from $M(y, z)$.

Important results were obtained from the projection $p(x, z)$, where, because of the c glide, the period along z is halved. An analysis of the peak distribution in $p(x, z)$ made it possible to separate all the peaks into two types. Each peak of the first kind with coordinates (x, z) corresponds to a peak at $(x + \frac{1}{2}, z)$, their heights differing by 10-15%. The peaks of the second kind, displaced by $(\frac{1}{2}, 0)$, have no mates. Such a peculiarity in the arrangement of the peaks, together with the fact that there is a strong peak at the point of symmetry $(\frac{1}{2}, 0)$ in the projection $p(x, z)$ (Fig. 1b), indicates the special position in the structure of seidozerite of two atoms with atomic numbers Z differing by 10-15%. The only special positions are the translationally nonidentical points of symmetry (projections of centers of symmetry or of two-fold axes) separated by $a/2$. We met with an analogous regularity in the analysis of the strucure of amblygonite [13] (and earlier still in the structure of epidote [17]).

The regularity in the Patterson functions of centrosymmetric crystals mentioned above [3] enabled us to find in $p(x, z)$ two maxima with coordinates (1.62, 0.36) and (0.55, 0.29) corresponding to distances between the atoms related by a point of symmetry. Besides these two vectors, a third $(\frac{1}{2}, 0)$ was used for the minimalization. The results of the minimalization from all three vectors were used in the construction of the

TABLE 1

Coordinates of the Atoms in the Asymmetric Unit of Seidozerite (in fractions of the axes a, b, c)

Atom	x	y	z	Atom	x	y	z
(4) Zr	0.200	0.119	0.074	(4) O_{II}	0.438	0.327	0,070
(2) Mn (Mg)	0.500	0.350	0.250	(4) O_{III}	0.438	0.907	0,071
(2) Ti	0.000	0,111	0.250	(4) O_{IV}	0,908	0.318	0.050
(4) Si_I	0.718	0.384	0.104	(4) O_V	0,915	0.912	0.056
(4) Si_{II}	0.718	0.843	0.104	(4) O_{VI}	0.804	0.314	0.191
(4) Na_I	0.195	0.611	0.069	(4) O_{VII}	0.804	0.915	0.192
(2) Na_{II}	0.000	0.613	0.250	(4) O_{VIII}	0.243	0,121	0.185
(2) Na_{III}	0.500	0.860	0.250	(4) (F,OH)	0.294	0.570	0.193
(4) O_I	0.719	0.615	0.105				

TABLE 2

Valency Balance in the Seidozerite Structure

Anion	Cation								Sum of valencies
	Si_I	Si_{II}	Zr	Ti	Mn	Na_I	Na_{II}	Na_{III}	
O_I	4/4	4/4							2
O_{II}	4/4		4/6			1/6			2 − 1/6
O_{III}		4/4	4/6			1/6			2 − 1/6
O_{IV}	4/4		4/6			2/6			2
O_V		4/4	8/6			1/6			2 + 1/2
O_{VI}	4/4			4/6	2/6		1/6		2 + 1/6
O_{VII}		4/4		4/6			1/6	1/6	2
O_{VIII}			4/6	4/6	2/6			1/6	2 − 1/6
(F,OH)					2/6	1/6	1/6	1/6	1 − 1/6

function M6 (x, z), which also gave the preliminary planar model of the structure. The refinement of the model was carried out by the subsequent calculation of the signs of the F_{h0l}'s and the calculation of the electron density projection ρ (x, z) (Fig. 2a).

One of the lines of symmetry in the (y, z) projection of the space group P2/c contains a glide, which facilitates the use of a statistical testing method [14].

The sign relationships of a basic group of reflections were found as the first stage of testing in the determination of the signs of the $0kl$ reflections of seidozerite, and the Zachariasen equation was systematically applied to find the signs of all the F_{0kl} as the second stage.

As is usual, the sign relationships arising from the structure factor formulae were used as a basis for the comparison of the amplitudes. For the $0kl$ zone of P2/c these relationships are as follows: 1) $l = 2n - F_{0kl} = F_{0\bar{k}l} = F_{0k\bar{l}} = F_{0\bar{k}\bar{l}}$; 2) $l = 2n + 1 - F_{0kl} = - F_{0\bar{k}l} = - F_{0k\bar{l}} = F_{0\bar{k}\bar{l}}$.

The 70 strongest unitary structure amplitudes were included out of the 331 nonzero $0kl$ in the basic group. A sign was considered as found if, in the general number of pairs being determined n > 3, the ratio of the number of pairs indicating the given sign to $\underline{n}$ was w ≥ 0.75 (for n = 3 it was necessary to take w = 1.0). Under these conditions the signs of 264 of the 331 nonzero $0kl$ were determined.

The electron density projection ρ (y, z) was constructed from the F_{0kl} with these "statistical" signs. In analyzing it we relied on the previously constructed ρ (x, z), the reliability of which was attested by the good agreement between the observed and calculated F_{h0l} .

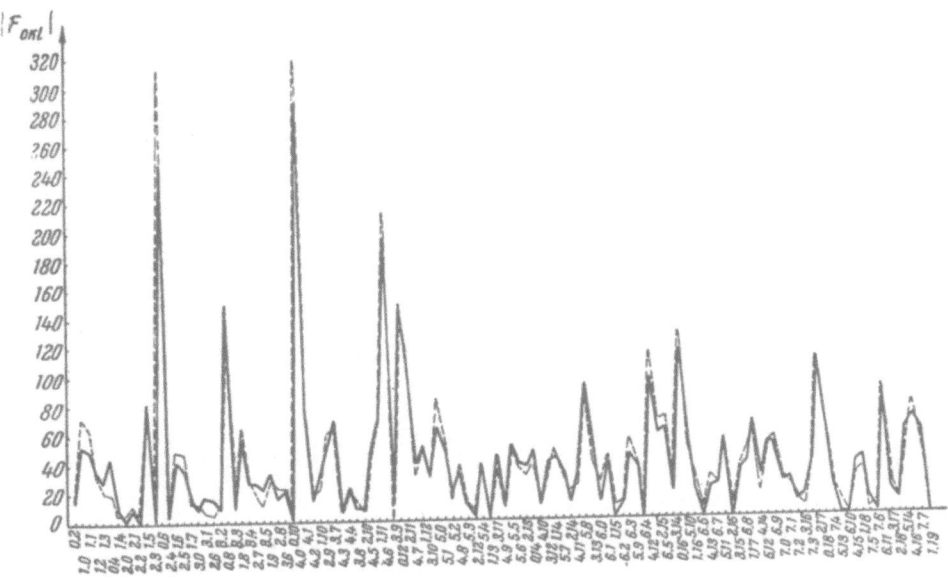

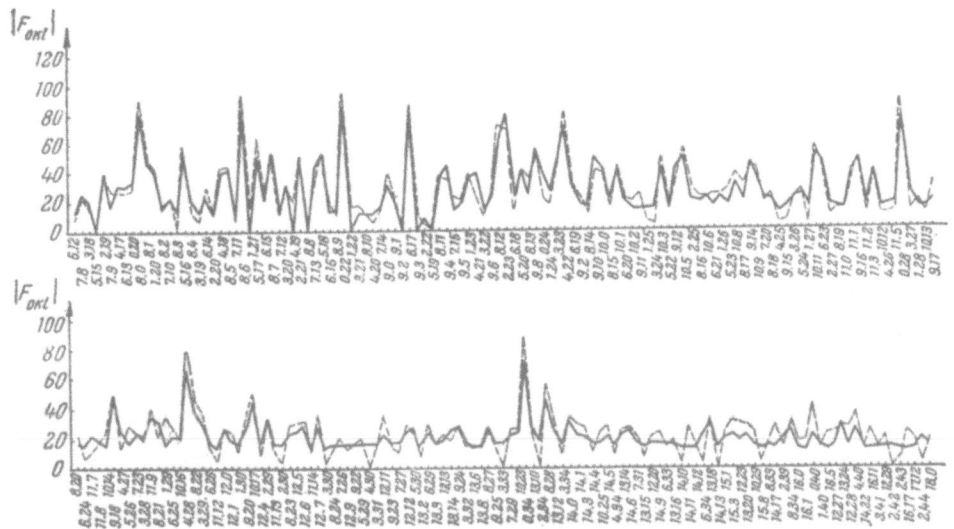

Fig. 3a. Graphical representation of the agreement between F_{obs} and F_{calc} for the $0kl$ zone.

The $\rho(y, z)$ synthesis showed a detrimental influence on the statistics of the geometrical regularity in the arrangement of the atoms in the (y, z) projection of seidozerite of which we spoke in the analysis of p(y, z). This regularity defeated the superpositions on p(y, z) and accounts for the fact that the statistics gave a large number (~ 25%) of wrong signs.

The presence in (y, z) of atoms lying at the points of an almost regular rectangular net b/4 × c/6, which appeared so clearly in p(y, z), is also visible in $\rho(y, z)$ (according to the statistical signs) with a strong heightening of the maxima on the two-fold axes.

Clearly, the most powerful $0kl$ reflections will be those for which the radiation scattered from the atoms lying at the points of the net b/4 × c/6 is in phase. It is just those reflections which were used in the basic group used in determining the signs of all the F_{0kl}. This is the reason why in $\rho(y, z)$ (calculated with the signs determined by the statistical method) maxima are lacking for the atoms which are not contained in the b/4 × c/6 net. The inclusion in the basic group only of reflections of strong unitary structure amplitude, representing a certain condensed subgroup of reflections, materially shortens the labor involved in using the testing method, but it leads to signs from which the electron density map shows only those atoms which give rise to the indicated

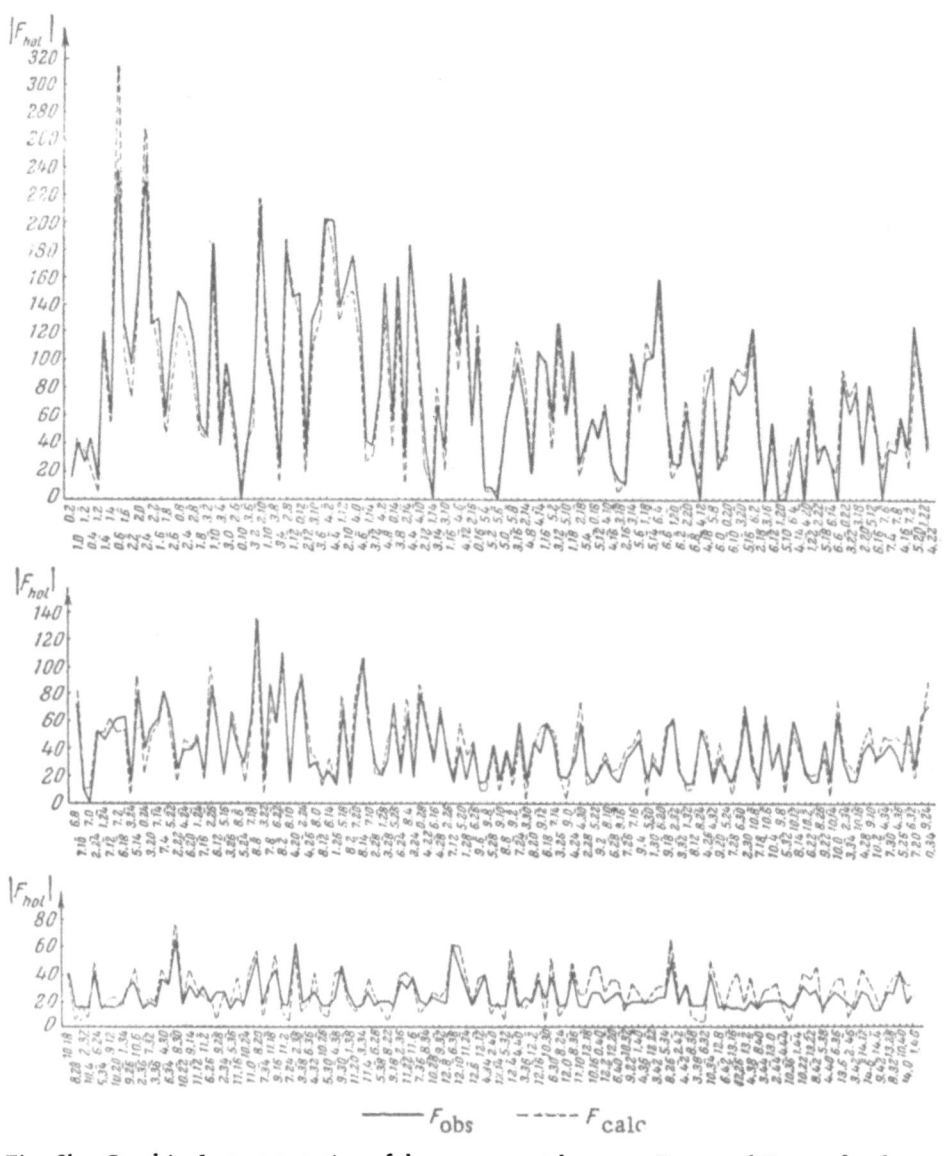

Fig. 3b. Graphical representation of the agreement between F_{obs} and F_{calc} for the h0l zone.

subgroup of reflections. Consequently the basic group must be such that it contains reflections whose contribution to the intensity from a given atom in the structure is important.

The y coordinates of the cations were determined from ρ (y, z) with the help of the known z coordinates from ρ (x, z) and the positions of the lighter anions were indicated so that it was possible to refine the projection by means of successive calculations of the signs of the F_{0kl} and of synthesis ρ (y, z).

The final electron density projections for seidozerite are shown in Fig. 2.

The interesting problem, from the point of view of crystal chemistry and mineralogy, of the isomorphous replacement in seidozerite had to be answered before the final variant of the structure could be selected. A comparison of the precise chemical analyses [1] with the heights of the maxima in the two electron density projections, and with the interatomic distances, led to the following picture of the isomorphous substitutions in seidozerite: 2Zr(1.44 Zr + 0.50 Ti); Mn(0.46 Mn + 0.34 Mg + 0.10 Fe··); Ti(0.76 Ti + 0.04 Nb + 0.10 Al + + 0.09 Fe···); 4 Na(3.58 Na + 0.38 Ca). The values shown in brackets to two significant figures are based primarily on the chemical analysis by Kazakova [1] and not on the heights of the electron density peaks.

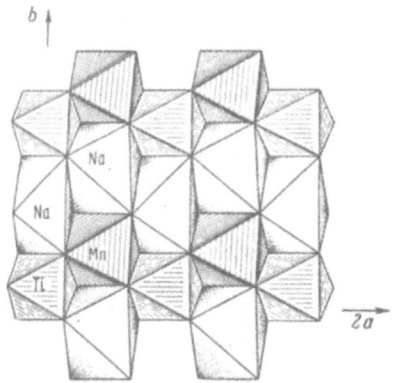

Fig. 4. Scheme showing the alternation of octahedra of different cations (Ti, Na $_{II}$, Mn, Na $_{III}$) in the continuous wall of seidozerite.

The effective atomic scattering functions were calculated, as general averages from the corresponding individual atom factors, for each of the crystallographically occupied positions. The F_{h0l} and F_{0kl} whose signs were used in the calculation of $\rho(x, z)$ and $\rho(y, z)$ where, in the first projection, the cell edges were divided into 60 parts along $\underline{a}$ and 120 along $\underline{c}$, and in the second both $\underline{b}$ and $\underline{c}$ were divided into 100 parts, were calculated using these f_{eff}.

The coordinates of the 17 atoms in the asymmetric unit of seidozerite (43 parameters), obtained by graphical interpolation, are given in Table 1.

The accuracy in the localization of the atoms from the projection was assessed, after Vainshtein [15], as: Zr ± 0.001 A, Mn ± ± 0.002 A, Ti ± 0.003 A, Si ± 0.004 A, Na ± 0.005 A, O ± 0.009 A. It must be mentioned that all the $\underline{x}$ coordinates were taken from $\rho(x, z)$ in which there are no single peaks. The pair similarity of the $\underline{x}$ coordinates for some of the overlapping atoms (Table 1) may not correspond to reality, but the symmetrical form of the peaks argues in favor of the reasonable closeness of the coordinates of the overlapping atoms. As the $\underline{a}$ axis is the shortest in seidozerite, an error in the relative $\underline{x}$ coordinate has less influence on the accuracy of the interatomic distances than one in the $\underline{y}$ or $\underline{z}$ coordinates.

F_{h0l} and F_{0kl} were calculated from the coordinates in Table 1 and f_{eff}, and a comparison of their absolute magnitudes with F_{obs} is shown in Figs. 3a and 3b. The discrepancies for the structure are, including all reflections $h0l$ and $0kl$ ($F_{obs} \neq 0$ to $\sin\theta/\lambda \leq 1.30$ and $F_{obs} = 0$ out to $\sin\theta/\lambda \leq 0.66$), are 17.2 and 22.5% respectively. Excluding the reflections with $F_{obs} = 0$ these values are 16.6 and 20.7%. The difference in the indexes for the two zones may be explained by the different effect of neglecting absorption on the $h0l$ and $0kl$ reflections.

The valency balance, according to Pauling's Second Rule (for ionic crystals) is given in Table 2. The greatest divergence in the sum of the valency strengths from the norm is observed for O_V. Formally, the excess of positive valencies is equal to 25% (a value which we may assume is the maximum possible) but in fact it is less because of the noticeably enlarged interatomic distance Zr–O_V* (Table 3).

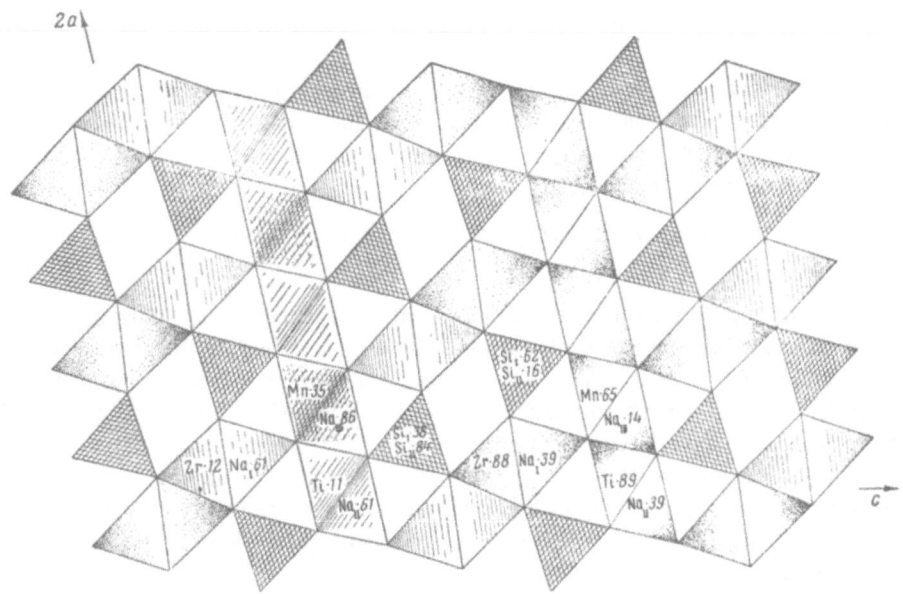

Fig. 5. Plan of the structure of seidozerite in coordination polyhedra. The (x, z) projection is perpendicular to the long axis of the crystal, the $\underline{y}$ coordinates of the cations are given in hundredths of the $\underline{b}$ axis.

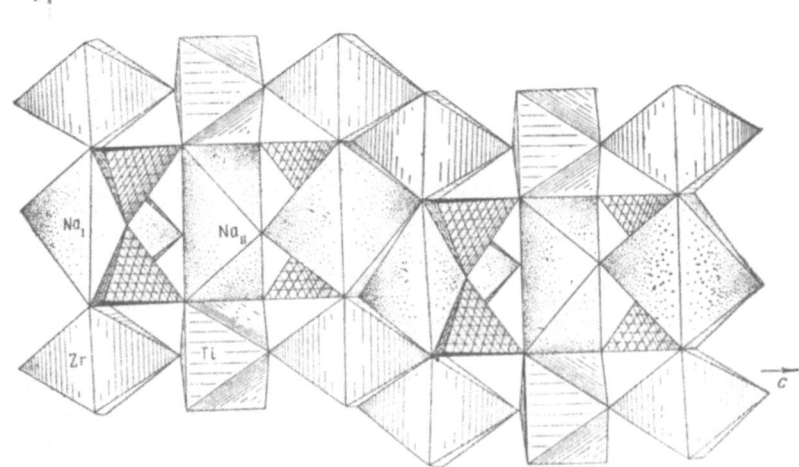

Fig. 6. The structure of seidozerite in coordination polyhedra. The (y, z) projection, omitting the columns of Mn-, Na_{III}-, Mn-, Na_{III}- octahedra which lie behind (in front of) the corresponding columns of Ti-, Na_{II}-, Ti-, Na_{II}- octahedra and are displaced relative to the latter by b/4 (Fig. 4).

III. Both the minerals from the Institute of Crystallography, ilvaite ($CaFe_2^{..}Fe^{...}O$- $[Si_2O_7]OH$) [16], epidote ($Ca_2Al_2FeO[Si_2O_7][SiO_4]OH$) and zorsite ($Ca_2Al_3O[Si_2O_7][SiO_4]OH$) [17, 18], and seidozerite, are, despite their orthosilicate stoichiometric formulae, diorthosilicates with $[Si_2O_7]$ groups and not $[SiO_4]$ groups. The eighth oxygen atom figuring in the chemical formula does not enter into the silicate radical.

In accordance with the structure found, the expanded formula of seidozerite must be written (doubling the stoichiometric formula) as $Na_4MnTi(Zr_{1.5}Ti_{0.5})O_2(F, OH)_2$ $[Si_2O_7]_2$, there being two such units in the cell.

The most characteristic feature of the architecture of seidozerite is the infinite compact walls of octahedra arranged along the two-fold axes. Cations of Mn(Mg), Ti and Na alternate in the octahedra according to the motif in Fig. 4. The parallel x0y planes of the wall are at a distance c/2 from one another and are connected to each other by infinite double ribbons extended along y and consisting of alternating Zr and Na octahedra. These architectural details are clearly visible in Fig. 5, which shows a plan of the structure in projection parallel to the long axis of the crystal. Two isolated diortho groups are each situated between the walls and the ribbons. Their axes are parallel to the y axis and the angle $Si_I - O_I - Si_{II}$ is 180° within the limits of accuracy of our measurements. The three intercepts uniting the corresponding O atoms of the upper and lower bases of the $[Si_2O_7]$ group are the edges of the three Na octahedra (Fig. 6), i.e., the diortho groups are situated not only between the walls and the ribbons, but between the Na octahedra.

In cuspidine $Ca_4[Si_2O_7]F_2^*$ and tilleyite $Ca_5[Si_2O_7](CO_3)_2$ [19] the diortho groups are linked in the same way with the Ca octahedra.

A further comparison of seidozerite with these Ca diorthosilicates is interesting. Tilleyite is made up of Ca octahedra in double (in width) infinite ribbons in two orientations. In cuspidine the analogously oriented ribbons of Ca octahedra are fourfold, and in seidozerite half the ribbons of one of the cuspidine orientations become infinite walls reinforced by tilleyite ribbons of the other orientation. The walls and ribbons are composed of columns in which the octahedra lie one on top of the other on edge, and in which, as is characteristic of a large number of Ca silicates [19], there are two octahedra in the 7.10 A b spacing. In tilleyite and cuspidine all the octahedra are occupied by Ca and in the ribbons and walls of seidozerite three kinds of Na octahedra alternate with Zr, Ti and Mn octahedra, and each time the Na_I, Na_{II} and Na_{III} octahedron, corresponding to the less tightly bound cations, is distorted, lengthening and shortening its edges so as most easily to enter the ribbon of octahedra round the more tightly bound atoms.

As was shown above, E. I. Semenov assigned seidozerite to the velerite group. According to the data given in the well known Mineralogical Tables of Strunz [20], these minerals are characterized by unit cell parameters which are similar to the parameters of cuspidine ($P2_1/c$, a = 7.53, b = 10.41, c = 10.83 A) [21]. The 7.19-7.33 A

*As in Original — Publisher's note.

TABLE 3

Interatomic Distances in the Seidozerite Structure in Å

Si_I — tetrahedron

Si_I — O_I	1.64
Si_I — O_{VI}	1.63
Si_I — O_{II}	1.62
Si_I — O_{IV}	1.66
O_I — O_{II}	2.56
O_I — O_{II}	2.65
O_{VI} — O_{II}	2.58
O_I — O_{IV}	2.58
O_{VI} — O_{IV}	2.52
O_I — O_{VI}	2.63
O_{IV} — O_{II}	2.70

Si_{II} — tetrahedron

Si_{II} — O_I	1.62
Si_{II} — O_{VII}	1.66
Si_{II} — O_{III}	1.60
Si_{II} — O_V	1.62
O_{III} — O_V	2.71
O_I — O_{VII}	2.64
O_I — O_{III}	2.58
O_I — O_V	2.63
O_{III} — O_{VII}	2.65
O_V — O_{VII}	2.69

Ti — octahedron

Ti — O_{VI}	1.98
Ti — O_{VII}	1.93
Ti — O_{VIII}	1.98
O_{VI} — O_{VIII}^{*}	2.71 (Mn)
O_{VI} — O_{VI}^{*}	2.71 (Na_{II})
O_{VII} — O_{VI}^{*}	2.68 (Na_{II})
O_{VIII} — O_{VII}	3.06 (Na_{III})
O_{VI} — O_{VII}	2.84
O_{VIII} — O_{VI}	2.82
O_{VIII} — O_{VIII}	2.86

Mn — octahedron

Mn — O_{VI}	2.21
Mn — O_{VIII}	2.31
Mn — (F,OH)	2.08
O_{VIII} — O_{VIII}^{*}	3.28 (Na_{III})
(F,OH) — (F,OH)*	2.73 (Na_{III})
O_{VIII} — (F, OH)	3.22
O_{VI} — O_{VIII}	3.37
O_{VI} — O_{VIII}^{*}	2.71 (Ti)
O_{VI} — (F,OH)	3.36
O_{VI} — (F,OH)*	2.95 (Na_{II})

Na_{II} — octahedron

Na_{II} — (F,OH)	2.14
Na_I — O_{VI}	2.52
Na_I — O_{VII}	2.51
O_{VI} — (F,OH)*	2.95 (Mn)
O_{VI} — O_{VI}^{*}	2.71 (Ti)
O_{VII} — O_{VII}	2.68 (Ti)
O_{VII} — (F,OH)*	3.34 (Na_{III})
O_{VI} — O_{VII}	4.26
O_{VI} — (F,OH)	3.26
O_{VII} — (F,OH)	3.65

Na_{III} — octahedron

Na_{III} — (F,OH)	2.48
Na_{III} — O_{VII}	2.21
Na_{III} — O_{VIII}	2.47
O_{VIII} — O_{VIII}	3.28 (Mn)
(F,OH) — (F,OH)*	2.73 (Mn)
O_{VIII} — O_{VII}	3.06 (Ti)
O_{VII} — (F,OH)*	3.34 (Na_{II})
O_{VIII} — (F,OH)	3.92
O_{VII} — (F,OH)	3.74
O_{VII} — O_{VIII}	3.41

TABLE 3 (Continued)

Zr — octahedron

Zr — O$_{VIII}$	1.98	
Zr — O$_{II}$	1.99	
Zr — O$_{IV}$	2.12	
Zr — O$_{III}$	2.01	
Zr — O$_V$•	2.13	
Zr — O$_V$•*	2.34	
O$_V$• — O$_V$*	2.73	(Zr)
O$_{II}$ — O$_{IV}$	2.87	(Na$_I$)
O$_{III}$ — O$_V$•	2.85	(Na$_I$)
O$_{IV}$ — O$_V$•	3.00	(Na$_I$)
O$_{IV}$ — O$_{III}$	2.98	
O$_V$• — O$_V$•	2.89	
O$_{II}$ — O$_V$•	3.16	
O$_{III}$ — O$_V$•*	2.98	
O$_{II}$ — O$_{VIII}$	2.96	
O$_{III}$ — O$_{VIII}$	2.97	
O$_{VIII}$ — O$_{IV}$	3.08	
O$_{VIII}$ — O$_V$	3.03	

Na$_I$ — octahedron

Na$_I$ — (F, OH)	2.24	
Na$_I$ — O$_{IV}$*	2.18	
Na$_I$ — O$_{II}$	2.42	
Na$_I$ — O$_{IV}$	2.60	
Na$_I$ — O$_{III}$	2.48	
Na$_I$ — O$_V$	2.62	
O$_{II}$ — O$_{IV}$	2.87	(Zr)
O$_{III}$ — O$_V$•	2.85	(Zr)
O$_{IV}$ — O$_V$•	3.00	(Zr)
O$_{IV}$ — O$_{IV}$*	3.45	(Na$_I$)
O$_{II}$ — O$_{III}$	4.12	
O$_{IV}$ — O$_{III}$	4.22	
O$_{II}$ — O$_{IV}$	3.46	
O$_{III}$ — O$_{IV}$	3.04	
O$_I$ — (F, OH)	3.08	
O$_{III}$ — (F, OH)	3.48	
O$_{IV}$ — (F, OH)	3.48	
O$_V$ — (F, OH)	3.78	

*The atoms indicated by an asterisk are linked with the corresponding atoms of the asymmetric unit by elements of symmetry. The cations with whose polyhedra the given edge is common are shown in brackets.

spacings in the minerals of the velerite group indicate the probable presence primarily of columns of (Ca, Zr, Mn) octahedra in their structures parallel to this translation and also of diortho groups with their axis along the same translation.

Whether the projections of these structures will be in a plane perpendicular to the axis of the diortho group, in an analogous manner to seidozerite or cuspidine, can only be revealed by further investigations. However, notwithstanding the resemblance of the chemical formulae of velerite, lovenite, guarinite and rosenbuschite on the one hand, and seidozerite on the other, the unit cells of the minerals of the velerite group correspond to the cell of cuspidine in the standard setting of the latter. This provides some reason for considering that the structure of the minerals of the velerite group is more nearly similar to the motif of cuspidine than to that of seidozerite.

A full list of interatomic distances in seidozerite is given in Table 3.

We will remark only upon those distances which differ most from the accepted ones. In the Zr octahedron, there are five Zr—O distances lying within the limits 1.98-2.13 A, and the sixth, Zr—O$_V$* is 2.34 A. As has been mentioned, this increase in the distance compensates for the excess of positive valencies accumulating at O$_V$ (Table 2). The edge O$_V$—O$_V$* = 2.73 A which is common to two Zr octahedra related by a center of symmetry is the shortest (among 12 O—O edges) in the Zr octahedron. In the present case, as will also be shown, some interatomic distances in the seidozerite structure provide good illustrations of Pauling's 3rd and 4th Rules.

In the Mn octahedron the Mn—(F, OH) distance of 2.08 A is also shortened, which may be considered also as compensating for the deficiency of positive charge on (F, OH) (5/6 instead of 1, Table 2). The edge O$_{VI}$ — O*$_{VIII}$ of the Mn octahedron also enters into the Ti octahedron, as a result of which it is shortened to 2.71 A, making it the shortest edge of the Mn octahedron. Further analysis of the interatomic distances on the basis of Table 3, in which those edges common to two polyhedra are specially marked, is not difficult.

The dimensions of the diortho group in seidozerite are in good agreement with the distances in the $[Si_2O_7]$ group in, for example, cuspidine [21]. The Si—O distances in seidozerite and cuspidine lie within the limits 1.60-1.66 A and 1.56-1.64 A respectively. The edges of the tetrahedra range from 2.52-2.71 A in seidozerite, and from 2.50-2.70 A in cuspidine.

Leaving aside the "anomalies" mentioned (and explained), then the cation—anion distances in the Zr, Ti, Mn and Na octahedra are close to the sum of the corresponding ionic radii. Ti—O are in the limits 1.93-1.98 A with the sum of the radii being $0.64 + 1.35 = 1.99$ A. Analogously for Mn—O they are 2.21-2.31 A with the sum $0.91 + 1.35 = 2.26$ A, for Na—O, 2.18-2.62 A with $0.98 + 1.35 = 2.33$ A (the shortest distance Na—(F, OH) = = 2.14 A). The sum of the ionic radii of Zr^{+4} and O^{-2}, $0.82 + 1.35 = 2.17$ A, and five of the six Zr—O distances lie in the range 1.98-2.13 A. The reduction in the effective value of these distances may be attributed to the displacement of about a fourth part of the atoms of Zr by Ti. The lengths of the edges of all the octahedra correspond to similar ones for the given cations found in other structures.

Let us say a word about the geometric properties of the seidozerite structure which led to the pecularities found in the beginning with the F and F^2 syntheses and which simplified the analysis. The fairly heavy concentrations of electrons which occupied the symmetry points separated by $a/2$ were not separate atoms but pairs of atoms, namely $Ti + Na_{II}$ and $Mn(Mg) + Na_{III}$; both these and others lie on the two-fold axes (Fig. 5). The ratio of the effective Z is $(22 + 12):(18 + 12) = 1.1$ and is equal to the difference in peak heights of 10-15% mentioned above. The $b/4 \times c/6$ net (Fig. 6) is made up of cations of $(Ti, Na_{II}, Mn, Na_{III})—(Zr, Na_I)—(Zr, Na_I)—(Ti, Na_{II}, Mn, Na_{III})—(Zr—Na_I)—(Zr—Na_I)$, arranged along the $\underline{c}$ axis at every 1/6th. The arrangement of the atoms at four levels perpendicular to $\underline{b}$ is shown in Fig. 2b.

The exact overlapping (in the xz projection) of the atoms Si_I, O_I, Si_{II}, and the triplet of O in the two bases of the Si tetrahedra is the result of the strict coincidence of the axis of the diortho group with the edges of the three Na octahedra. By the same token, the $[Si_2O_7]$ group registers a true trigonal prism.

E. I. Semenov found perfect cleavage along (001) in seidozerite. This plane is parallel to the continuous walls of octahedra (Figs. 5, 6) along which passes the cleavage in various silicates. This same cleavage may be caused by the breaking of bonds along the (001) plane which cuts the $\underline{c}$ axis in half (Fig. 5). In this case only the longer $Zr—O^*_V = 2.34$ A bonds and the generally weaker Na—O bonds will be broken.

Seidozerite is an optically biaxial mineral and is positive, $2V = + 68°$, $n_g = 1.830$, $n_m = 1.758$, $n_p = 1.725$ (according to E. I. Semenov). The major axis of the optic indicatrix coincides with $\underline{a}$, and the minor with $\underline{b}$, and the explanation of this may be seen in the fact that the most optically active cations, the Mn, join, in their octahedra along $\underline{a}$, with the "strong" (+ 4) Ti in zigzag chains, whilst along $\underline{b}$ the Mn octahedra alternate with the extended octahedra around the weak (loosely bound, with a charge of + 1) Na cation (Fig. 4).

LITERATURE CITED

[1] E. I. Semenov, M. E. Kazakova and V. I. Simonov, Zapiski Vsesoyuz. Mineral. Obshchestva 87, 5, 590 (1958).

[2] E. R. Howells, D. C. Phillips and D. Rogers, Acta Cryst. 3, 216 (1950).

[3] Kh. S. Mamedov and N. V. Belov, Doklady Akad. Nauk SSSR 106, 462 (1956).

[4] A. E. Fersman, Selected Works. 3 [In Russian] (1955) p. 146.

[5] K. A. Vlasov et al., Geochemistry of the Lovozerskii Massif [In Russian] (Izd. AN SSSR, Moscow, 1959).

[6] A. G. Betekhtin, Mineralogiya [In Russian] 1950, p. 719.

[7] A. I. Kitaigorodskii, X ray Structural Analysis [In Russian] (1950) p. 185.

[8] M. J. Buerger, Proc. Nat. Acad. Sci. (Wash) 36, 376 (1950).

[9] M. J. Buerger, Proc. Nat. Acad. Sci. (Wash) 36, 738 (1950).

[10] I. D. Thomas and D. McLachlan, Acta Cryst. 5, 301 (1952).

[11] B. K. Vainshtein, Trudy Inst. Krist. 7, 15 (1952).

[12] M. J. Buerger, Proc. Nat. Acad. Sci. (Wash) 39, 678 (1953).

[13] V. I. Simonov and N. V. Belov, Kristallografiya 3, 4, 428 (1958).*

[14] I. M. Rumanova, Trudy Inst. Krist. 10, 59 (1954).

[15] B. K. Vainshtein, J. Éxptl.-Theoret. Phys. (USSR) 27, 44 (1954).

[16] N. V. Belov and V. I. Mokeeva, Trudy Inst. Krist. 9, (1954).

[17] N. V. Belov and I. M. Rumanova, Trudy Inst. Krist. 9, 103 (1954).

[18] E. G. Fesenko, I. M. Rumanova and N. V. Belov, Kristallografiya 1, 2, (1956).

[19] N. V. Belov, Min. sborn. L'vov. geol. Obshchestva 10, 14 (1956).

[20] H. Strunz, Mineralogische Tabellen (1957) p. 273.

[21] R. F. Simonova, I. M. Rumanova and N. V. Belov, Zapiski Vsesoyuz. Mineral. Obshchestva 84, 159 (1955).

Received January 21, 1959

---

*[Soviet Physics — Crystallography, p. 429].

Reprinted from Doklady of the Academy of Sciences of the USSR (Doklady Akademii Nauk SSSR),
Earth Sciences Sections, Vol. 126, pp. 574-576, May-June, 1959

THE CRYSTALLINE STRUCTURE OF TRICALCIUM SILICATE HYDRATE*

$$TSH = 6CaO \cdot 2SiO_2 \cdot 3H_2O = Ca_6[Si_2O_7](OH)_6 = Ca_4[Si_2O_7](OH)_2 \cdot 2Ca(OH)_2.$$

Kh. S. Mamedov, R. F. Klevtsova, and Academician N. V. Belov

The determination [1] of the structure of cuspidine $Ca_4[Si_2O_7]$ (F, OH)$_2$ immediately after the structure [2] of tilleyite $Ca_5[Si_2O_7](CO_3)_2 = Ca_4 Si_2O_7 CO_3 \cdot CaCO_3$ was an essential preliminary step towards the identification of a number of Ca silicates, in particular wollastonite and xonotlite [3].

The basic characteristic of Ca silicates, as opposed to Mg(Fe), Al silicates investigated considerably earlier (with regard to structure) was clearly distinguished for the first time in cuspidine and tilleyite; i.e., the role which [SiO$_4$] tetrahedrons play in Mg(Fe) and Al silicates is occupied by [Si$_2$O$_7$] groups in Ca silicates. The causes are purely geometrical. The edges of the [SiO$_4$] tetrahedrons (2.7-2.8A) are not commensurate with the edges of the Mg, Fe, Al octahedrons (3.6-3.7A). The heights of the [Si$_2$O$_7$] groups are commensurate with the latter, and this makes it almost inevitable that these groups be introduced in the Ca silicates either singly or as (clearly distinguishable) links in more complex silica radicals (wollastonite, xonotlite, apophyllite, feldspars).

The somewhat clumsy character of the [Si$_2$O$_7$] groups and the presence of only one direction in them with the indicated value of 3.7A (height of the group) substantially limit the variety of the motifs in the composition of which the [Si$_2$O$_7$] appears in Ca silicates. Thus, in both the Ca silicates mentioned at the beginning of this article the "tilleyite band" appears in the role of a mineralogical radical in a characteristic manner. In the projection (Fig. 1, 1) we see that its recurrent increment, the wave, is composed of four Ca octahedrons and two Si tetrahedrons. In Fig. 1, 1, we see that each Ca octahedron is the face of an infinite column of Ca octahedrons standing on the edges by which the Ca octahedrons are mutually connected above and below. As regards the Si tetrahedrons, they are connected in pairs into Si$_2$O$_7$ groups which themselves acquire the form of trigonal prisms of 6 O atoms in which the group Si-O-Si is included. These prisms are inscribed in wider and infinite rectilinear trigonal channels with walls - double bands of Ca columns - of such a character that each edge of a prism is at the same time the edge of three neighboring Ca octahedrons. Here, however, a layer of Ca octahedrons with a sandwiched [Si$_2$O$_7$] group alternates with a layer of Ca octahedrons without an [Si$_2$O$_7$] group, along the axis of the channel.

As may be seen from Fig. 1, 1, the tilleyite band link, occupying two stages along the a axis, is formed by eight Ca octahedrons and two [Si$_2$O$_7$] groups. The 14 O atoms from the two [Si$_2$O$_7$] groups included in the formula $Ca_8[Si_2O_7]_2$ are insufficient to balance the charges of the

cations; this insufficiency is compensated by additional (according to V. S. Sobolev [4]) anions (F, OH) in the structure of cuspidine, which is entirely composed of tilleyite bands with the very simple formula

$$Ca_8[Si_2O_7]_2 (F, OH)_4 = 2Ca_4 [Si_2O_7] (F, OH)_2.$$

Of the two simplest geometrical solutions for such a formula (Fig. 1, 2, and 3) the second variant is found in nature; in this variant, by extending one another in the neighboring bands the links formed out of Ca octahedrons of a tilleyite band produce somewhat longer 4-membered links which - since they overlap each other- make the structure stronger in appearance. The tilleyite bands do not merge with the structure of tilleyite itself but retain their individuality and are combined with each other by additional [CO$_3$] groups; this also requires the introduction of an additional Ca octahedron and the formula will, therefore, be $Ca_4[Si_2O_7]CO_3 \cdot CaCO_3$ (Fig. 2).

It is easy to calculate that in two single-stage layers along the x axis in cuspidine there are 8 atoms of O (F) at the apexes of the Ca octahedrons (half the 9th atom in the center of the Si-O-Si group is not included in this number). Two [CO$_3$] groups give 6 additional O in exchange for 2 F(OH), i.e., the total number of O(F) atoms is increased by one-and-a-half times (12:8). Whereas in cuspidine there are four layers of Ca octahedrons along the x axis, in tilleyite there are six and the parameter a is correspondingly one and a half times greater, as may be seen from the data of Table 1. The ratio of the a axes is somewhat less than 1½ there, as a result of the well-known contradiction of the O-O edges in the [CO$_3$] group from 2.7 to 2.2A.

In 1958 Buckle, Gard, and Taylor published new data on the hydrothermal synthesis of tricalcium silicate hydrate (TSH), a product of great importance in the chemistry of cements; they consider that the following formula may be safely adopted for this compound:

$$3CaO \cdot SiO_2 \cdot 1.5H_2O$$
or
$$6CaO \cdot 2SiO_2 \cdot 3H_2O = Ca_6[Si_2O_7](OH)_6.$$

By electron diffraction methods (using minimum single crystals) and powder diagrams the authors obtained the parameters of the rhombic cell given in Table 1. In view of the general structural regularities

---

*Translation of: O Kristallicheskoi Strukture Gidrata Trekhal'tsievogo Silicatas, Doklady Akademii Nauk SSSR, vol. 126, no. 1, 1959, pp. 151-154.

of Ca hydrosilicates together with these data and the chemical analysis, it may be assumed that we have a further instance of a representative of a structure with tilleyite bands and, moreover, one which is closer to tilleyite than in the case of cuspidine. The dimensions of $\underline{b}$ and $\underline{c}$ are similar in all three Ca silicates. The dimensions of $\underline{a}$, which is somewhat greater in TSH than in tilleyite, makes it possible to assume that here also we have an instance of a six-layer structure.

If we deduct the formula of the tilleyite band, i.e., the formula of cuspidine from that of tricalcium silicate hydrate:

$$Ca_6 [Si_2O_7] (OH)_6 - Ca_4[Si_2O_7] (OH)_2 = 2Ca (OH)_2,$$

we obtain two molecules of portlandite $Ca(OH)_2$ which must enter the formula in the "side chain" outside the cuspidine nucleus (Table 1). With their 4 OH molecules, they form a third layer which links (two-layer) tilleyite bands and contains additional Ca octahedrons (Fig. 3), replacing the CO_3 groups in tilleyite.

The absence of $[CO_3]$ groups removes the cause of the reduction of the $a_{til} : a_{cus}$ ratio to a value less than 1.5. On the contrary, it is now somewhat greater than 1.5 because the other cause which had reduced $\underline{a}$ in cuspidine, namely (Fig. 1,1) the possibility of the drawing-in of the corresponding Ca octahedrons from the neighboring layer along the $\underline{a}$ axis by the third apex of the SiO_4 tetrahedrons, is absent. Correspondingly, in TSH we have a volume of $26.2A^3$ per O atom compared with $22.2A^3$ in cuspidine and only $21.6A^3$ in tilleyite.*

The $[Si_2O_7]$ groups in TSH are found on the axes of the perfectly regular triangular empty channels and are surrounded by six columns - three (double) bands of Ca

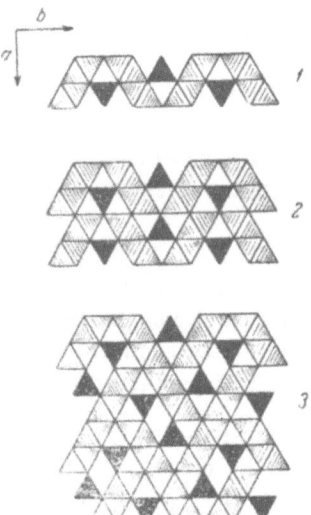

Fig. 1. Structure of cuspidine $Ca_4 [Si_2O_7]$ (F, OH)$_2$; 1) (tilleyite) Band with the indicated structure, the condensation of which gives cuspidine; 2) simplest variant of the cuspidine structure; 3) actual variant.

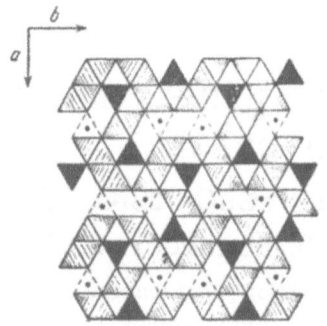

Fig. 2. Structure of tilleyite Ca$_5$ $[Si_2O_7] (CO_3)_2$; shaded triangles with a dot in the center signify a $[CO_3]$ group.

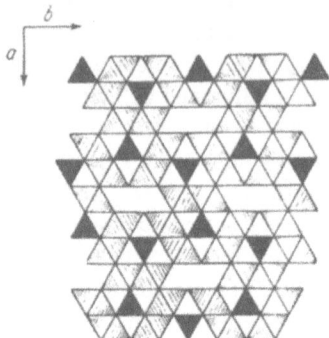

Fig. 3. Structure of TSH; the tilleyite bands are linked by pairs of Ca octahedrons.

octahedrons. This makes it possible to regard the TSH structure as pseudohexagonal and similar to the anhydrous tricalcium silicate. But it is hardly possible to speak of true hexagonality of the crystals, as is claimed by the authors of the work [5], who carried out an experimental investigation. They insist on the "hexagonal" ratio a: :b = √3 = 1.78 and in order to obtain this ratio they use the parameter $\underline{a}$ from one sample and the parameter $\underline{b}$ from another; both their samples, however, give the same value a:b = 1.65. Another argument in favor of hexagonality - the centered character of the plane C = ab (characterizing the hexagonal structure in the "ortho-hexagonal" system) -indicates only a tendency to a centered character, which is readily seen in Fig. 3. This only weakens the spots with h + k ≠ 2n, just as in all fibrous Ca silicates, spots of 00l with $l \neq$ 2n are weakened (this is readily seen in the electron diffraction patterns reproduced in the article by the three authors mentioned above); as a result, along the $\underline{c}$ axis we have two Ca octahedrons and only one $[Si_2O_7]$ group. The pseudo-hexagonal cleavage surfaces $10\bar{1}0$ and $11\bar{2}0$ indicated by these authors are prominent in our system (in the rhombic symbols 100 and 210).

As a result of the essential role played by the tilleyite band, a corresponding radical for the description of

---

*The first number approximates the value we have for the structure of CaO (27.2 A^3).

TABLE 1

Mineral		a	b	c	z	$V_O(\text{Å}^3)$
Cuspidine $Ca_4[Si_2O_7](F, OH)_2$	C_{2h}^5	10.41	10.15	7.55	4	22.2
Tilleyite $Ca_5[Si_2O_7](CO_3)_2 = Ca_4[Si_2O_7]CO_3 \cdot CaCO_3$	C_{2h}^5	14.46	10.27	7.50	4	21.8
Tricalcium silicate hydrate $Ca_6[Si_2O_7](OH)_6 = Ca_4[Si_2O_7](OH)_2 \cdot 2Ca(OH)_2$	D_{2h}	17.2	10.5	7.63	4	26.2

Note. In the table certain changes have been made to the original data. First, the indices of the axes have been unified. The indices of the spatial groups are given in a general form because the monoclinic axis in cuspidine is the (a) axis perpendicular to the tilleyite bands, whereas in tilleyite the (b) axis coincides with the direction of the band. A rectangular (pseudo) cell has been selected in all cases and monoclinic angles are, therefore, absent in the table. In actual fact, the value of b given for cuspidine is $b \cos \alpha$ and in the same way the value of a given for tilleyite is $a \cos \beta$.

crystalline structure and for establishing models of new structures (other examples, apart from TSH see in [6]), it would be desirable to distinguish this band - roughly speaking the cuspidine "molecule" - when writing the formulas of minerals; this is done in Table 1.

The thermogram of tricalcium silicate hydrate given by the three authors shows that the moisture is liberated in two stages separated by a short but definite break. In the first stage somewhat more than two thirds is liberated, the remainder being liberated in the second stage. The loss in weight ends at ~700°. On the X-ray pattern, samples corresponding to heating to 500-600° give γ-Ca_2SiO_4 lines which, however, remain anomalous up to 700°, at which point CaO lines, hitherto absent, are added to it.

This behavior of TSH is in complete agreement with that which should be expected on the basis of the formula written in the expanded form with the tilleyite band (cuspidine "molecule") distinguished. Undoubtedly, the two OH groups included in this molecule are more firmly linked than the four OH in the "side chain." The two stages of dehydration may be represented by the equations:

$$Ca_4[Si_2O_7](OH)_2 \cdot 2Ca(OH)_2 \rightarrow |Ca_4[Si_2O_7](OH)_2| + 2CaO$$
$$+ 2H_2O \uparrow \rightarrow 2\gamma\text{-}Ca_2SiO_4 + H_2O \uparrow + 2CaO.$$

The tendency of CaO to a finely dispersed state in the presence of H_2O is due to the very late appearance of the CaO lines compared with diorthosilicate molecules, which also increase slowly.

Received February 26, 1959

REFERENCES

1. Smirnova, R. F. et al., Zap. Miner. Obshch, vol. 84, no. 2, 159, 1955.
2. Smith, J. V., Acta Crystal., vol. 6, 9, 1953.
3. Mamedov, Kh. S., and N. V. Belov, Doklady Akademii Nauk SSSR, vol. 104, p. 615, 1955; vol. 107, p. 463, 1956.
4. Sobolev, V. S., VVEDENIE V MINERALOGIYU SILIKATOV [INTRODUCTION TO THE MINERALOG' OF SILICATES]; L'vov, 1949.
5. Buckle, E. R., J. A. Gard, and H. F. W. Taylor, J. Chem. Soc., p. 1351, 1958.
6. Mamedov, Kh. S., V. I. Simonov, and N. V. Belov, Doklady Akademii Nauk SSSR, vol. 126, no. 2, 1959.

Reprinted from Soviet Physics—Crystallography, Vol. 4, pp. 300-314, March, 1960

THE STRUCTURES OF HERDERITE, DATOLITE AND GADOLINITE DETERMINED BY DIRECT METHODS

P. V. Pavlov and N. V. Belov

Institute of Crystallography, Academy of Sciences of the USSR
Lobachevskii University, Gorkii

(Translated from: Kristallografiya, Vol. 4, No. 3, pp. 324-340)

(Original article submitted February 23, 1959)

The direct methods developed at the Institute of Crystallography and at Gorkii University have proved exceptionally powerful in solving the structures of the three minerals named in the heading; we give a detailed exposition of the solution with illustrations of the use of Harker-Kasper inequalities and of statistical analysis by Zachariasen's methods.

Herderite $CaBePO_4F$ [1-3] is of special interest because it contains beryllium (15.3 wt. % BeO, as against 14% BeO in beryl). Strunz [4] has supposed that herderite has a structure very like that of datolite $CaBSiO_4(OH)$, which Ito and Mori have examined [5] by the (half) heavy-atom method; the two should form a good example of how ions of equal size but different valency, having the same sum of all valencies for the groups (pairs), may replace one another to give the same structure:

$$CaBSiO_4 (OH)$$

$$CaBePO_4F.$$

Gadolinite also contains $Fe^{..}$: datolite is $Ca_2B_2(OH)_2Si_2O_8$, while gadolinite is $Fe^{..} Y_2Be_2O_2Si_2O_8$.

The heredite crystal (from the Mineralogical Museum, Academy of Sciences of the USSR) had a largest dimension of about 1 mm; it had no faces, so the axes were found from Laue patterns by the Umanskii-Kvitka method [6]. The rotation patterns were used to confirm Strunz's parameters $a = 9.80$ A, $b = 7.68$ A, $c = 4.80$ A, $\beta = 90°06'$, $a:b:c = 1.276:1:0.625$ and the space group $C_{2h}^5 = P2_1/a$. This group is one of the commonest for organic and inorganic substances, and is convenient in that it contains a center of symmetry. The density (3.00 g/cm^3) and volume of the cell show that the cell contains four $CaBePO_4F$ units (molecules). The four pairs of centers of symmetry (duplicated positions without parameters) are the only points in the cell that are not in general positions with three parameters.

The x-ray patterns included three rotation photographs and Weissenberg photographs of the equatorial layer lines rotated about $\underline{a}$, $\underline{b}$, and $\underline{c}$, and of the 1st, 2nd, 3rd and 4th layer lines rotated about $\underline{c}$. The Mo-radiation produced many reflections, which are of value when direct methods of either kind are used.

Mamedov has used Harker-Kasper inequalities with Zachariasen's statistical methods to determine the signs of the reflections from the oscillation photograph for one layer line only in the case of $6CaSiO_3 \cdot H_2O = Ca_6[Si_6O_{17}](OH)_2$

(monoclinic) [7]. Rumanova [8] has developed similar methods for use with epidote (monoclinic). These methods have been much simplified to give the signs of the amplitudes for herderite.

The xy projection corresponds to the shortest repeat period (4.8 A). The $\underline{c}$ oscillation photograph had 167 independent hk0 reflections not equal to 0, whose intensities were estimated from the blackening [9]. The amplitudes were referred to the absolute scale (as is needed if the Harker-Kasper inequalities are to be used) by Wilson's [11] and Vainshtein's [12, 13] methods. Both methods consider only the squares of the amplitudes and need no preliminary model. Wilson's method is statistical and demands reflection derived from the whole volume of the reciprocal lattice. We followed Mamedov in using only hk0 reflection, which is equivalent (see [14]) to assuming that the mean intensities are the same for all layers. The reflections, including the zero F_{hk0}, were related to the $\sin\vartheta/\lambda$ values by division into intervals 0.1-0.2, 0.2-0.3, 0.3-0.4, etc., i.e., into concentric rings with $(\sin\vartheta_1/\lambda - \sin\vartheta_2/\lambda) = 0.1$ in the equatorial section of the reciprocal lattice. All the reflections within a ring were used to find the means $\overline{F^2_{hk0}}$ and the sums of the squares of the mean atomic factors $\sum_i \overline{f_i}^2$ from

$$\overline{F^2_{hk0}} : \sum_i \overline{f_i^2} = K^2 \exp\left(-2B \sin^2\vartheta/\lambda^2\right),$$

where K is the factor for passing from the experimental amplitudes to the absolute ones. The curves $\log \Sigma \overline{f_i^2}/\overline{F^2_{hk0}}$ vs $\sin^2\vartheta/\lambda^2$ gave K^2 as 184, i.e., K = 13.6, for B = 0.58. We also used only the hk0 reflections in Vainshtein's method:

$$\frac{1}{S}\sum |F_{hk0}|^2 = \sum_i g_i',$$

where S = (ab) is the area of that face of the cell. We must be given B and must allow for overlap between atoms in order to find K from this formula. Vainshtein's table [13] shows that B does not exceed one for herderite. We may expect that overlap is improbable along the very short axis c = 4.8 A; because the mean intensities of layer lines 0-4 are almost equal, the Vainshtein's formula has $g'_i = 0.38 \cdot Z^{2.37}$ with B = 1 without allowance for overlap.

The calculated g' and $\Sigma|F_{hk0}|^2$ gave $K^2 = 194$, i.e., that K = 13.9. The values agree well enough for us to assume that no exceptional errors are present.

The next step is to find the absolute unit structure amplitudes from

$$|U_{hk0}| = |F_{hk0}| : \sum_i f_i,$$

where f_i is the atomic factor of the i-th atom. These $|U_{hk0}|$ included 15 with values > 0.5, so in establishing the signs we could use the simplest inequality:

$$|U_{hkl}|^2 \leqslant \frac{1}{2} + \frac{1}{2} U_{2h, 2k, 2l}, \tag{1}$$

which gives the result quickly when $|U_{2h, 2k, 2l}|$ and $|U_{hkl}|$ are both larger than 0.5. But the only reflection found at once to have a positive sign was 24.0.0 ($U_{24.0.0} = 0.65$ and $U_{12.0.0} = 0.50$).

To find the signs of the other hk0 reflections, we used the basic inequalities·

$$(U_H + U_K)^2 \leqslant (1 + U_{H+K})(1 + U_{H-K}), \tag{2}$$

$$(U_H + U_K)^2 \leqslant (1 - U_{H+K})(1 - U_{H-K}), \tag{3}$$

where H represents (hkl), etc..

We pick out amplitudes with $|U| \geq 0.45$ and use a simplified reciprocal grid to select many $|U_{hk0}|$ triplets for which (2) and (3) give the sign relations $S_{H \pm K} = S_H \times S_K$, where S is the sign of an amplitude. Let us consider some particular examples of the use of (2) and (3). In the list of unit amplitudes we find:

$$|U_{640}| = 0.64; \quad |U_{10.\bar{3}.0}| = 0.66; \quad |U_{16.1.0}| = 0.63; \quad |U_{\bar{4}70}| = 0.49.$$

The corresponding inequalities are

$$(U_{640} + U_{10.\bar{3}.0})^2 \leq (1 + U_{16.1.0}) \cdot (1 + U_{\bar{4}70}),$$ (2')
$$(U_{640} - U_{10.\bar{3}.0})^2 \leq (1 - U_{16.1.0}) \cdot (1 - U_{\bar{4}70}).$$ (3')

The four possible values of the RHS for the four meaningful combinations of the signs to $U_{16.1.0}$ and $U_{\bar{4}70}$ are

$$(1 + 0.63) \cdot (1 + 0.49) = 2.43$$
$$(1 + 0.63) \cdot (1 - 0.49) = 0.83$$
$$(1 - 0.63) \cdot (1 + 0.49) = 0.55$$
$$(1 - 0.63) \cdot (1 - 0.49) = 0.19.$$

The signs of the two items on the left must be either the same or different. If $S_{640} = S_{10.\bar{3}.0}$, the LHS in (2') is $(0.64 + 0.66)^2 = 1.69$. The inequality is satisfied if the RHS is 2.43. For this to be so we must have $S_{16.1.0} = S_{\bar{4}70} = +$, i.e., we must have both positive. If $S_{640} = -S_{10.\bar{3}.0}$, we see from (3') that the LHS is still 1.69, but that on the RHS the signs of the 16.1.0 and $\bar{4}70$ reflections must both be minus, i.e., that $S_{16.1.0} = S_{\bar{4}70} = -$. We cannot say which of these two contrary results is correct, since we do not know whether S_{640} and $S_{10.\bar{3}.0}$ have the same sign or different signs, though both give the consistent result

$$S_{16.1.0} = S_{\bar{4}70} = S_{640} \cdot S_{10.\bar{3}.0},$$

i.e., 16.1.0 and $\bar{4}70$ have the same sign, which is the product of the signs of 640 and $10.\bar{3}.0$. This expresses the fact that 16.1.0 and $\bar{4}70$ have the same sign if 640 and $10.\bar{3}.0$ also have the same sign.

Consider a case in which the three amplitudes are large (≥ 0.5) and the fourth is nearly zero:

$$|U_{14.3.0}| = 0.67; \quad |U_{800}| = 0.51; \quad |U_{22.3.0}| = 0.57; \quad |U_{630}| = 0.07.$$

Again we have

$$(U_{14.3.0} + U_{800})^2 \leq (1 + U_{22.3.0}) \cdot (1 + U_{630}),$$ (2")
$$(U_{14.3.0} - U_{800})^2 \leq (1 - U_{22.3.0}) \cdot (1 - U_{630}).$$ (3")

The four possible values for the RHS are

$$(1 + 0.57) \cdot (1 + 0.07) = 1.68$$
$$(1 + 0.57) \cdot (1 - 0.07) = 1.46$$
$$(1 - 0.57) \cdot (1 + 0.07) = 0.46$$
$$(1 - 0.57) \cdot (1 - 0.07) = 0.40.$$

The maximum value for the LHS is $(0.67 + 0.51)^2 = 1.39$.

Again we suppose that $S_{14.3.0} = S_{800}$. Then (2") is satisfied if the RHS is 1.68, i.e., if $S_{22.3.0} = S_{630} = +$ (a). But it is satisfied also if the RHS is 1.46, which is possible if $S_{22.3.0} = +$, $S_{630} = -$ (b).

If we compare (a) and (b), we see that the sign of 630 stays undefined. If now we suppose that $S_{14.3.0} = -S_{800}$, (3") then gives us that $S_{22.3.0} = S_{630} = -$ when the RHS is 1.68, and that $S_{22.3.0} = -$, $S_{630} = +$ when the RHS is 1.46, i.e., S_{630} is still undefined. Thus, we conclude that $S_{22.3.0} = S_{14.3.0} \times S_{800}$, and that we can draw no conclusion about the sign of 630.

Interesting examples occur when the signs of two of the reflections are related symmetrically. Thus we have

$$|U_{640}| = 0.64; \quad |U_{\bar{6}40}| = 0.64; \quad |U_{12.0.0}| = 0.50; \quad |U_{080}| = 0.39.$$

The symmetry of group P2₁/ a implies that the signs of amplitudes related by the mirror plane must be the same when h + k = 2n, i.e., that $S_{640} \cdot S_{\overline{6}40} = +1$. We use (2) directly. This is satisfied when the LHS is 1.64 and the RHS is 2.08, whereupon $S_{080} = S_{12.0.0} = +$, i.e., both signs are plus. It would be contrary to the symmetry of the group to suppose that 640 and $\overline{6}40$ have different signs. Thus,

$$S_{080} = S_{12.0.0} = S_{640} \cdot S_{\overline{6}40} = +.$$

Now for h + k = 2n + 1 the symmetry of P2₁/ a demands that hk0 and h$\overline{k}$0 have different signs, i.e., that reflections from planes related by the mirror plane should have different signs (and hence equal moduli). For example, $|U_{10.3.0}| = 0.66$; $|U_{10.\overline{3}.0}| = 0.66$; $|U_{20.0.0}| = 0.44$; $|U_{060}| = 0.45$. Then we have $S_{10.\overline{3}.0} = -S_{10.3.0}$.

But (3) is satisfied if the LHS is 1.74 and the RHS is 2.09, i.e., if

$$S_{20.0.0} = S_{060} = -.$$

The symmetry of the group prevents 10.3.0 and 10.$\overline{3}$.0 from having the same sign. These relations between the signs, dependent on the symmetry, are of great assistance in finding the signs. It is clear that some inequalities give us either only the relations between the signs of two reflections or (if both indices are even) that the sign is plus. The relations implied by the symmetry can give also a minus sign (again, only for reflections whose two indices are even).

These useful ways of finding the sign relations can be to a large extent mechanized with the graphs given in [15]. In all we found 45 confirmations of the sign relations $S_{H \pm K} = S_H \cdot S_K$ for the following sets of three:

$(\overline{10}.11.0)$	22.$\overline{3}$.0	16.$\overline{7}$.0×640		5.13.0	$\overline{5}$.10.0×10.3.0
$(\overline{4}10)$	16.7.0	10.3.0×640		20.0.0	800×12.0.0
$(\overline{8}10)$	20.7.0	14.3.0×640		22.3.0	800×14.3.0
$(\overline{5}\overline{5}0)$	17.13.0	11.9.0×640	$(10.10.0)$	18.$\overline{4}$.0	$4\overline{7}0$×14.3.0
$(\overline{6}40)$	18.4.0	12.0.0×640	$(\overline{6}20)$	6.10.0	640×060
$(5\overline{8}0)$	7.16.0	1.12.0×640	$(\overline{12}.6.0)$	12.6.0	12.0.0×060
$(\overline{10}.5.0)$	22.3.0	16.$\overline{1}$.0×640		10.3.0	10.$\overline{3}$.0×060
$(\overline{4}70)$	16.1.0	10.$\overline{3}$.0×640		5.16.0	5.10.0×060
(080)	12.0.0	$6\overline{4}0$×640		$\overline{5}$.16.0	$\overline{6}40$×1.12.0
$(\overline{12}.8.0)$	24.0.0	18.$\overline{4}$.0×640		$\overline{11}$.12.0	12.0.0×1.12.0
	18.$\overline{4}$.0	12.$\overline{8}$.0×640	$(\overline{2}30)$	10.11.0	640×470
$(0.14.0)$	12.$\overline{6}$.0	6.$\overline{10}$.0×640	$(\overline{8}70)$	16.7.0	12.0.0×470
$(1.14.0)$	11.$\overline{6}$.0	5.$\overline{10}$.0×640		10.3.0	$6\overline{4}0$×470
	5.16.0	$\overline{1}$.12.0×640		800	470×$4\overline{7}0$
$(\overline{5}70)$	15.13.0	10.3.0×5.10.0		13.11.0	10.3.0×380
$(17.\overline{5}.0)$	$\overline{5}$.13.0	$\overline{11}$.9.0×640	$(\overline{4}.13.0)$	16.7.0	10.$\overline{3}$.0×6.10.0
$(\overline{11}.11.0)$	17.5.0	14.$\overline{3}$.0×380		1.14.0	$6\overline{4}0$×7.10.0
$(\overline{11}.11.0)$	21.9.0	16.$\overline{1}$.0×5.10.0	$(\overline{2}40)$	14.4.0	800×640
	7.16.0	$\overline{4}70$×11.9.0	$(6.10.0)$	18.2.0	12.6.0×$6\overline{4}0$
$(\overline{2}30)$	22.3.0	12.0.0×10.3.0		21.9.0	15.5.0×640
$(\overline{4}60)$	24.0.0	14.$\overline{3}$.0×10.3.0		15.5.0	640×990
(060)	20.0.0	10.$\overline{3}$.0×10.3.0		19.$\overline{2}$.0	15.5.0×$4\overline{7}0$
$(14.\overline{4}.0)$	6.10.0	$4\overline{7}0$×10.3.0			

If the signs S_H and S_K are denoted by different letters in these relations, the sign of $S_{H \pm K}$ is the product of those letters. The inequalities enabled us to express many of the letters (whose number initially was rather large) in terms of products of two letters; ultimately there remained only a, b, and ab. Thus, we compiled a reference group from the 56 reflections (33%) which had signs. The sign of any amplitude having both indices even could be found directly. The simplest case was that of 24.0.0 ; the signs of 18 other reflections of this type were deduced in a more complicated way. All amplitudes having h = 2m, k = 2n + 1 had signs denoted by a, whereas those having h = 2m, k = 2n had ones denoted by ab, and those having h = 2m+1, k = 2n + 1, had ones of b.

85

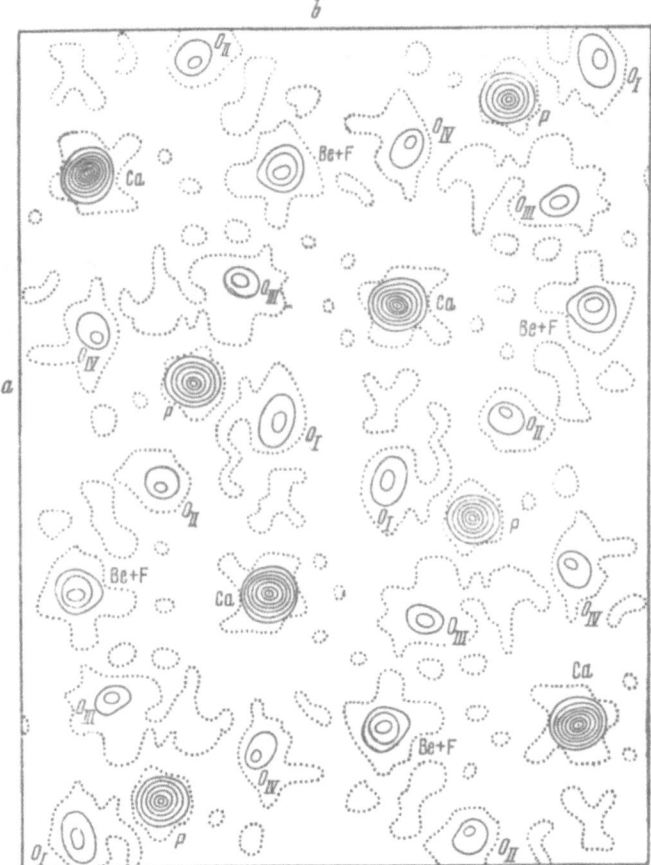

Fig. 1. The xy projection of the electron-density contours for herderite.

In a primitive lattice we may [8] assign arbitrarily three signs for the three-dimensional case, or two for a projection, subject to the condition that these signs are given to reflections falling in different groups. We have only two letters, so we may make them both plus.

The signs of the 56 reference reflections were checked by Zachariasen's method [16] and were used to find the signs of the other F_{hk0} by purely statistical methods. This process was successful for 84 amplitudes, i.e., of the 167 signs we established 140 with a "probability" (ratio of number of pairs implying that sign to the total number of pairs) not less than 70%. The number of defining pairs per F_{hk0} was 20-24.

The electron-density pattern constructed from the 140 (of the 167) F_{hk0} gave a well-resolved picture with large peaks, which we naturally equated to Ca atoms; the next most important peak fell at the centers of alternating squares and triangles formed by the less obvious peaks. The distances between the atoms corresponding to these peaks were such as to indicate that the squares were projections of the PO_4 tetrahedra, and that the triangles were projections of BeO_4 tetrahedra standing on their bases; the central Be atom is covered by the fourth O (of F). The xy projection of the electron density agrees well with the corresponding Patterson diagram, which we constructed intially, but which gave no useful results. The divergence coefficient (R) at this stage was 24%, and only two of the F_{hk0} calculated from the atomic coordinates had signs the reverse of those given by the direct method. A second xy projection, which included the remaining 27 F_{hk0} (with calculated signs), gave R = 14.9% without the zero reflections, and 20.4% with them (for all F_{hk0} up to sin ϑ/λ = 1.1); Fig. 1 shows this second projection; Fig. 2 compares the calculated and experimental structure amplitudes. The experimental values were numberd in accordance with the condition $\Sigma F_0 = \Sigma F_T$, which showed that the previous deduction of the absolute values was in error by not more than about 5%, i.e., could not interfere with the inequalities.

The z coordinates were deduced from the xz projection, which also checked the x and y. The oscillation photograph on b for the equatorial layer had 116 non-zero h01, all with h = 2n on account of the glide plane, which halves the x axis. The h values for all reflections have been divided by 2 to conform to the smaller cell. The signs were again found by using the inequalities and statistical methods. There were 21 reflections with $|U_{h0\ell}| \geq 0.5$, so conditions were even more favorable; we were able to show that the relation $S_{H \pm K} = S_H \cdot S_K$ was correct for the following 60 sets of amplitudes:

	109	105×004		1200	$90\bar{3} \times 303$
	404	400×004	$(\bar{3}09)$	$90\bar{3}$	$60\bar{6} \times 303$
$(\bar{6}04)$	604	600×004		602	$30\bar{1} \times 303$
$(\bar{9}03)$	905	901×004		409	400×009
	$\bar{3}05$	$\bar{3}01 \times 004$	(006)	600	$30\bar{3} \times 303$
$(30\bar{1})$	$\bar{3}09$	305×004	(008)	$60\bar{2}$	$30\bar{5} \times 303$
(602)	$\bar{6}06$	$\bar{6}02 \times 004$		301	$00\bar{4} \times 303$
	703	$70\bar{1} \times 004$	$(\bar{2}02)$	10.02	602×400
(303)	309	$\bar{3}03 \times 006$	$(\bar{2}04)$	10.04	604×400
(606)	$\bar{6}06$	$\bar{6}00 \times 006$		$90\bar{3}$	$50\bar{3} \times 400$
$(\bar{6}08)$	604	$60\bar{2} \times 006$		$90\bar{5}$	$50\bar{5} \times 400$
	705	$70\bar{1} \times 006$		$70\bar{1}$	$30\bar{1} \times 400$
	404	$40\bar{2} \times 006$	(105)	$70\bar{5}$	$30\bar{5} \times 400$
(309)	$\bar{3}03$	$\bar{3}03 \times 006$	$(\bar{2}06)$	10.02	$60\bar{2} \times 404$
(506)	$\bar{5}.0.12$	$\bar{5}03 \times 009$		703	$\bar{3}01 \times 404$
(304)	$\bar{3}.0.14$	$\bar{3}05 \times 009$	(109)	$70\bar{1}$	$30\bar{5} \times 404$
$(\bar{5}05)$	705	105×600		705	$30\bar{1} \times 404$
	$\bar{2}08$	$\bar{3}03 \times 105$	(004)	$12.0\bar{4}$	$60\bar{4} \times 600$
	$\bar{3}05$	400×105	(303)	$90\bar{3}$	$30\bar{3} \times 600$
	404	$30\bar{1} \times 105$	(305)	$90\bar{5}$	$30\bar{5} \times 600$
	602	$50\bar{3} \times 105$	(901)	303	$30\bar{1} \times 602$
$(\bar{5}07)$	703	$60\bar{2} \times 105$	(1200)	004	$602 \times 60\bar{2}$
$(\bar{3}01)$	905	602×303	(307)	$90\bar{3}$	$30\bar{5} \times 602$
	703	400×303		$50\bar{3}$	$\bar{1}05 \times 602$
$(\bar{3}05)$	901	$60\bar{2} \times 303$	$(60\bar{2})$	$6.0.\bar{1}0$	006×604
	$12.0.\bar{4}$	$50\bar{3} \times 701$	$(90\bar{7})$	303	$\bar{3}05 \times 60\bar{2}$
	404	$\bar{3}01 \times 703$		$12.0.\bar{4}$	$50\bar{5} \times 707$
(406)	10.04	$30\bar{1} \times 705$		$3\bar{0}9$	$00\bar{9} \times 300$
	901	004×903		$12.0.\bar{4}$	$60\bar{2} \times \bar{6}06$
(301)	$90\bar{5}$	$30\bar{3} \times 60\bar{2}$		12.0.0	$50\bar{5} \times 705$

TABLE 1

Atomic Coordinates for Herderite, Datolite, and Gadolinite (as hundredths of the a, b and c axes). Datolite I gives the Japanese Results; Datolite II Gives our Data

Atoms	Herderite			Datolite I			Datolite II			Gadolinite	
	x	y	z	x	y	z	x	y	z	x	y
Ca — Y	33 0	11.3	99.5	33.8	10.3	99.0	33.9	10.8	99.0	32.7	10.8
P — Si	7.9	26.9	47.5	8.3	26.5	47.5	8.6	26.2	47.1	7.7	28.3
Be — B	33.3	40.8	55.8	34.0	40.0	60.0	34.0	41.0	57.6	33.4	40.5
O_I	4.1	39.7	25.0	4.0	39.5	25.0	4.0	39.5	24.2	2.1	39.6
O_{II}	45.6	27.8	65.0	46.0	30.0	69.0	46.0	30.4	67.1	44.4	28.2
O_{III}	19.5	34.3	67.3	21.0	33.0	70.0	21.0	32.7	67.6	20.1	33.4
O_{IV}	14.1	11.1	33.3	14.0	8.5	32.0	14.2	9.2	32.0	15.7	10.7
F — OH	33.0	41.7	20.9	33.5	41.5	27.5	33.6	41.5	25.8	33.5	40.5
Fe··	—	—	—	—	—	—	—	—	—	0	0

But here the result was to leave three letters instead of two (a, b, and c) and the products ab, ac, bc, and abc. Many (14) of the amplitudes with both indices even were shown to be positive, but about half of this type were represented by the letter b. The xy projection enabled us to choose between b = + and b = − because it

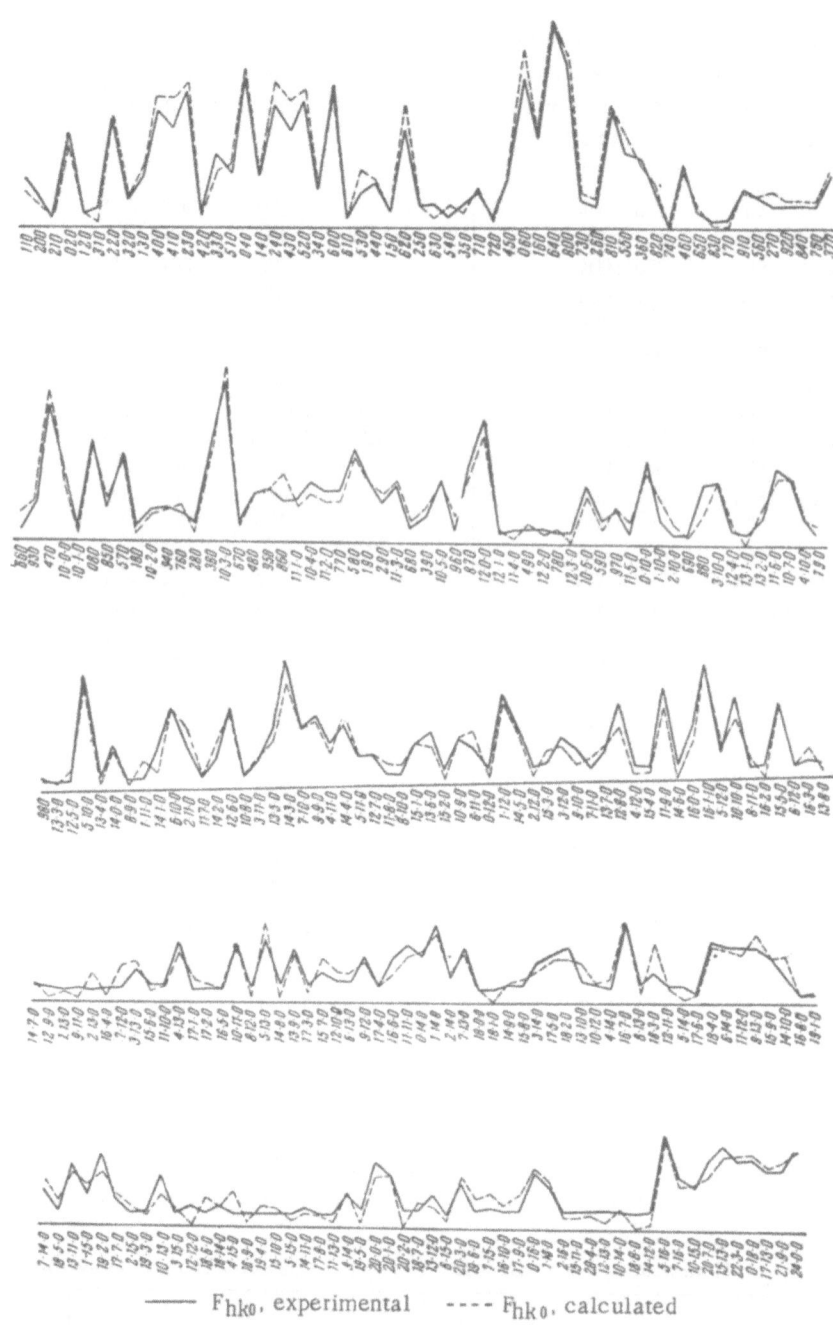

Fig. 2. Agreement between the calculated and experimental $|F_{hk0}|$ for herderite.

had good R (20.4%). The sign of the common reflection 800 was negative in that projection, whereas in the xz one the sign of 800 (400) was b. Hence b = −*, and the other signs are defined by a, c, and ac. If we had solved the xz projection independently, we could have chosen a and c arbitrarily (e.g. could have made them plus), but we can choose only three signs arbitrarily [8], and two had been used up in the xy projection; so if we were to put a = c = +, we had to ensure that this agreed with the first projection, as in the case of b. The appropriate reflection for a and c is 300 (600), which has a plus sign in xy, and here has the sign ac; hence a and c must have the same sign. We put a = c = + and took ten h00 amplitudes whose signs had been defined in xy, and thereby

_____

* This could be expected a priori, since in any other case all the signs of the h = 2m, k = 2n group would be plus.

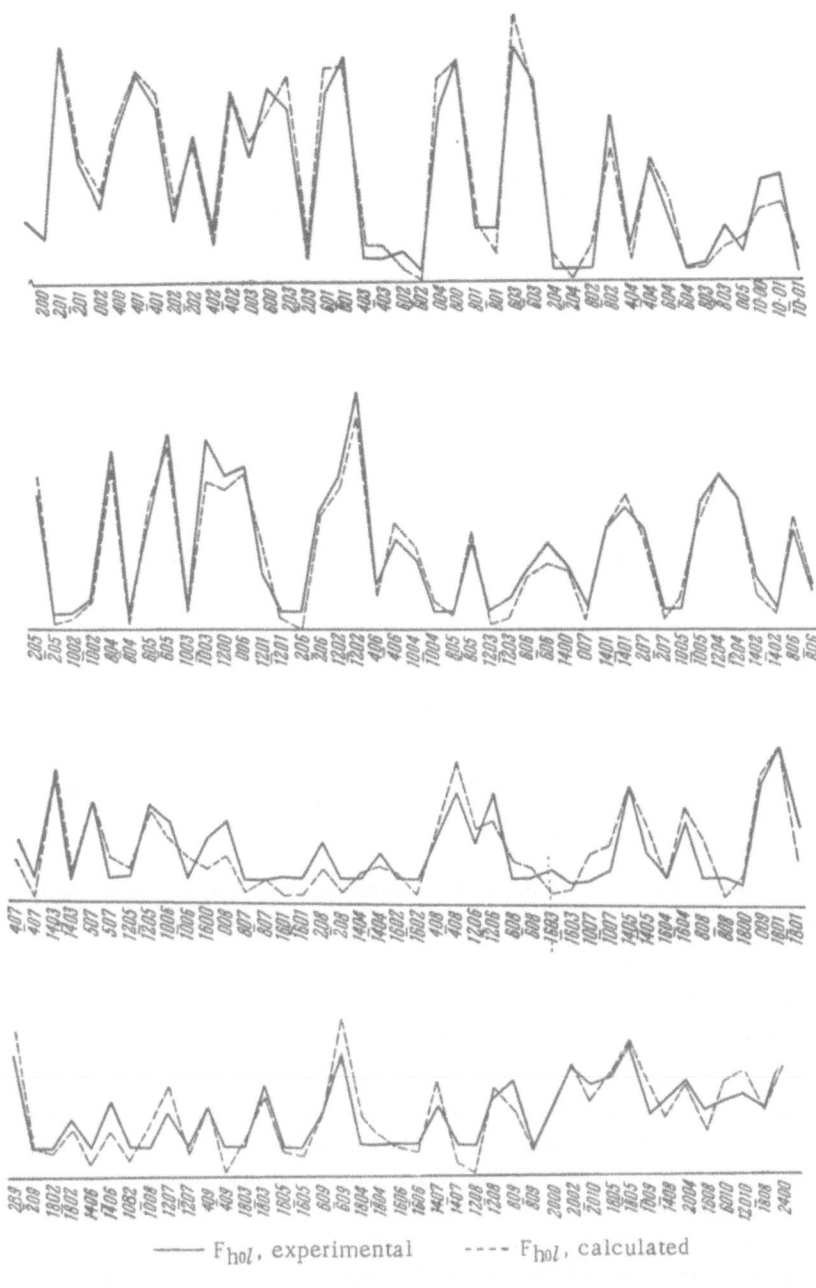

Fig. 3. Agreement between the calculated and experimental $|F_{h0l}|$ for herderite.

obtained a reference group of 61 reflections. Statistical methods with numbers of defining pairs up to 16 gave the rest of the signs. The mode was as before; the probability was 100% for the minimum number of pairs (3) and was not less than 70% when the number was large. There were 26 signs left in xz, which were not given by the first projection; these were calculated. The convergence coefficient for the repeated xz projection was 14.3% without the experimental zero reflections, or 19.7% with the 60 zero reflections, again for all F_{h0l} up to $\sin\vartheta/\lambda = 1.1$. Figure 3 shows how well the theoretical and experimental $|F_{h0l}|$ agree.

The xz projection (Fig. 4) shows few peaks; but they are sharp, being the Be atoms we took to be covered by O(F) in the xy projection.

Again, we checked the coordinates from the oscillation photograph on a (9.80 A) for the zero layer which had 67 nonzero reflections. The y and z coordinates from the xy and xz projections were used to calculated the

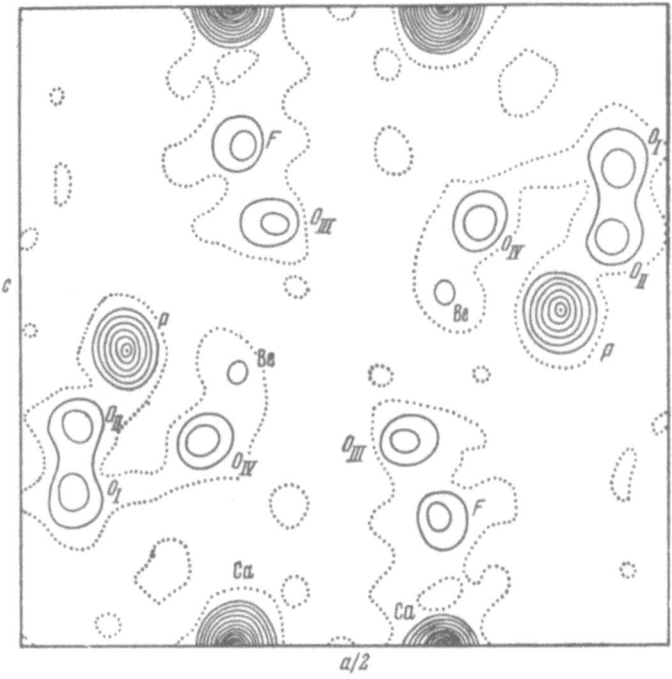

Fig. 4. The xz electron-density contours in projection for herderite.

F_{0kl}, which were compared with the experimental values. Here R = 17.4% for the 67 nonzero reflections and 19% for 83 reflections with 16 zeros up to ϑ/λ = 0.85

Table 1 gives the coordinates derived from the xy and xz projections. The atoms all lie in fourfold position, i.e., there are 24 independent parameters for the eight atoms in a formula unit.*

We used Vainshtein's method [13] to calculate the probable error in the coordinates by using the theoretical and experimental F_{hkl}; the value for Ca was 0.004 A; for P, 0.006 A; for Be, 0.033 A; for O, 0.015 A; and for F, 0.013 A. As usual, these errors are determined by the errors made in estimating the intensities, and are increased by the errors introduced by interpolation.

Figure 5 shows that herderite contains infinite pseudotetragonal nets, consisting of PO_4 and BeO_3F tetrahedra, which species have different orientations. The nets consist of centrosymmetric rings made up from the two species of tetrahedron, which rings alternate with closed centrosymmetric octagons made up from the same two species. This pattern occurs in feldspar and in other closely allied aluminosilicates [17]. In Fig. 5 the atoms are given their z coordinates as hundredths of c; we see that the structure has two layers. Figure 6 shows the two layers separately as Pauling's polyhedra. The lower layer is made of slightly distorted cubes (Ca), which are typical of Ca in garnet, or of Cu in $CuAl_2$ [18]. The upper layer is a pseudotetragonal network of PO_4 tetrahedra (with one orientation) and of BeO_3F tetrahedra (with another).

Table 2 gives the interatomic distances. The P − O distances do not fall outside the range 1.51-1.57 A, while the O − O edges in the PO_4 tetrahedron range from 2.43 to 2.56 A. Other distances are: Be− O = 1.55-1.67 A and Be− F = 1.67 A, with O − O(F) = 2.52 −2.73 A. The eight distances Ca−O = 2.40 -2.70 A, most of them being close to the lower figure. The PO_4 tetrahedra are much smaller than SiO_4 ones, but the BeO_3F tetrahedra are almost the same size as SiO_4 ones.

---

* The x coordinates have been reduced by 0.50a relative to those found by assigning the signs arbitrarily, which has been allowed for in the xy projection (Fig. 4) and in calculating the theoretical F. The origin has been moved in order to facilitate comparing our results with Ito's. This displacement is used in all following illustrations.

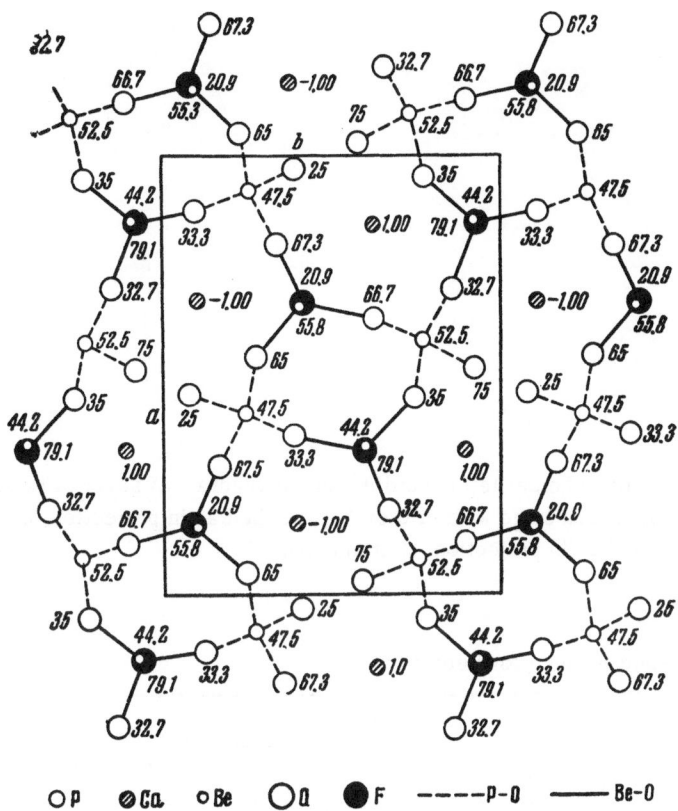

Fig. 5. The xy projection of the structure of herderite. The numbers are the atomic coordinates in hundredths of the axis c = 4.80 A.

TABLE 2

Atomic Distances in Herderite (A)

Tetrahedron		Tetrahedron		Polydedron	
$P — O_I$	1.51	$Be — O_{II}$	1.64	$Ca — O_I^*$	2.40
$P — O_{II}^{**}$	1.52	$Be — O_{III}$	1.55	$Ca — O_I^*$	2.40
$P — O_{III}$	1.57	$Be — O_{IV}^*$	1.67	$Ca — O_{II}$	2.44
$P — O_{IV}$	1.51	$Be — F$	1.67	$Ca — O_{III}$	2.61
$O_I — O_{II}^{**}$	2.49	$O_{II} — O_{III}$	2.52	$Ca — O_{III}^*$	2.70
$O_I — O_{III}$	2.56	$O_{II} — O_{IV}^*$	2.73	$Ca — O_{IV}$	2.44
$O_I — O_{IV}$	2.43	$O_{II} -- F$	2.68	$Ca — F$	2.71
$O_{II}^{**} — O_{III}$	2.51	$O_{III} — O_{IV}^*$	2.62	$Ca — F^{**}$	2.58
$O_{II}^{**} — O_{IV}$	2.53	$O_{III} — F$	2.64		
$O_{III} — O_{IV}$	2.46	$O_{IV} — F$	2.67		

---

* Atoms related to the corresponding basic ones by the symmetry elements of the group.

** Atoms in adjacent cells.

Table 3 gives the sums of the valences compiled in accordance with Pauling's second rule for the common O-corners of the polyhedra. The balances are not satisfied exactly for O_I and O_{III}, but the lack of 1/4 at O_I and the same excess at O_{III} account for only 12.5% of the oxygen valences.

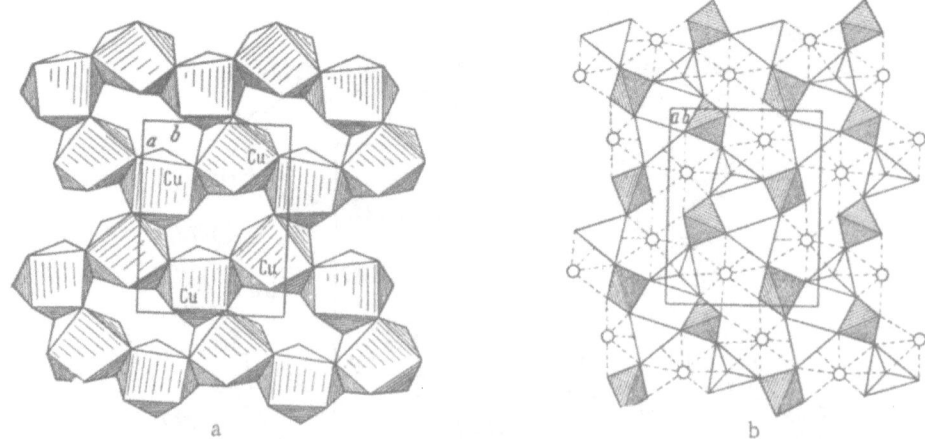

Fig. 6. The structure of herderite in coordination polyhedra. a) Layer of twisted cubes with Ca atoms; b) second stage with grid of PO_4 tetrahedra (in projection, as squares) and of BeO_3F tetrahedra (in projection, as triangles).

T A B L E 3

Valence Balance for Herderite

Anion	Cation			Sum of valences
	Ca	Be	P	
O_I	$2 \cdot 1/4$	—	$1 \cdot 5/4$	$7/4 = 2 - 1/4$
O_{II}	$1 \cdot 1/4$	$1 \cdot 1/2$	$1 \cdot 5/4$	2
O_{III}	$2 \cdot 1/4$	$1 \cdot 1/2$	$1 \cdot 5/4$	$9/4 = 2 + 1/4$
O_{IV}	$1 \cdot 1/4$	$1 \cdot 1/2$	$1 \cdot 5/2$	2
F	$2 \cdot 1/4$	$1 \cdot 1/2$	—	1
			Check	9.0

The structure of datolite $CaBSiO_4(OH)$ (a = 9.62, b = 7.60, c = 4.84 A; β = 90°; same space group) has been given by Ito and Mori. They used the xy Patterson projection to find the $\underline{x}$ and $\underline{y}$ coordinates of the Ca and Si atoms, from which they calculated the signs of the F_{hk0}. They used the 30 strongest F_{hk0} to construct the xy projection of the electron density, which revealed, in addition to the known Ca and Si, three peaks due to O atoms, which confirmed 55 signs as reliable; then a second synthesis was made, and then from 95 F_{hk0}, a third. The resulting coordinates were taken as final. The $\underline{z}$ coordinates were deduced from the chemical crystallography of the substance and were refined from a Fourier projection of the xy plane; the $\underline{z}$ coordinate of the B atom, which overlaps OH, was found from a section taken along [001].

The structure shows certain differences from that of herderite, especially in relation to the coordinates of the Ca atom (Table 1). In datolite the $\underline{y}$ for Ca is 10.3, whereas in herderite it is 11.3; the $\underline{z}$ for B is 60, whereas the one for the analogous Be in herderite is 55.8; the $\underline{z}$ values for O_{II} and O_{III} are 69 and 70 and 65 and 67.3, respectively. In datolite OH has z = 27.5, whereas in herderite F has z = 20.9. We have revised the coordinates of datolite by determining the structure afresh by the direct methods presented above.

For the xy projection we had 158 nonzero F_{hk0} (Mo radiation), as against 95 in the Japanese work. Wilson's method gave K = 13.3, and all the amplitudes were transferred to the absolute scale with B = 0.5 and were calculated to $|U_{hk0}|$ form; then they included 12 which were greater than 0.5. Then (2) and (3) were used to confirm that $S_{H \pm K} = S_H \cdot S_K$ was true for the sets listed below, which were selected from those with $|U_{hk0}| \geq 0.40$.

	22.4.0	14.3̄.0×81̄0		24.0.0	10.3̄.0×14.3.0
	20.7.0	12.6̄.0×81̄0		14.3̄.0	060×14.3.0
	16.0.0	81̄0×81̄0	(18.4̄.0)	10.10.0	4̄70×14.3.0
	24.0.0	16.1̄.0×81̄0	(16.7.0)	4̄.13.0	1̄0.3.0×6.10.0
	14.3̄.0	64̄0×81̄0	(16.10.0)	1̄6.10.0	16.0.0×0.10.0
	12.6̄.0	47̄0×81̄0		1̄.17.0	7.13.0×640
	10.10.0	2.1̄1̄.0×81̄0		1.1̄2.0	7.1̄3.0×81̄0
	6.1̄0.0	2.1̄1̄.0×81̄0		11.12.0	6.10.0×520
(12.6̄.0)	12.6.0	06̄0×12.0.0	(5̄80)	15.12.0	10.10.0×520
	6.10.0	6̄.10.0×12.0.0		7.13.0	2.11.0×520
(230)	22.3.0	12.0.0×10.3̄.0		15.5.0	10.3.0×520
(16.1̄.0)	470	6̄40×10.3̄.0		1.12.0	6.10.0×5̄20
	6.10.0	47̄0×10.3̄.0	(21.9.0)	1̄.15.0	1̄1.12.0×10.3.0
	10.3̄.0	06̄0×10.3̄.0		990	470×520
	22.3̄.0	12.6̄.0×10.3̄.0		6̄.10.0	1̄2.6.0×640
	6̄40	47̄0×10.3̄.0	(16.1.0)	4̄70	1̄0.3.0×640
(62̄0)	6.10.0	060×640		6̄40	1̄2.0.0×640
(080)	12.0.0	64̄0×640		16.1̄.0	470×12.6.0
(87̄0)	20.1.0	14.3̄.0×640		24.0.0	12.6̄.0×12.6.0
(2.11.0)	10.3.0	47̄0×640	(6̄20)	18.10.0	12.6.0×640
(0.14.0)	12.6̄.0	6.1̄0.0×640	(4̄60)	12.8.0	81̄0×470
	640	6̄40×12.0.0		580	060×520
(10.4̄.0)	22.4.0	640×16.0.0		10.10.0	6.10.0×400
(12.7̄.0)	20.7.0	470×16.0.0	(4̄10)	4.13.0	470×060
(81̄0)	20.7.0	640×14.3.0	(1̄6.6.0)	6.16.0	16.0.0×06̄0
	10.10.0	6̄.10.0×16.0.0	(2̄90)	14.11.0	8.1.0×6.10.0
	24.0.0	81̄0×16.1.0	(800)	0.14.0	470×470
	22.3̄.0	64̄0×16.1.0			

The signs of all those amplitudes whose two indices are positive were determined directly; those with h = 2m + 1, k = 2n were denoted by $\underline{c}$, those with h = 2m, k = 2n + 1 by $\underline{a}$, and those with both odd, by ac. Both letters were assumed to be plus. The inequalities gave the signs of 47 reflections, which gave us a reference group from which to find the other F_{hk0} by statistical methods; this gave us the signs of a further 90 reflections. The probabilities were as before with numbers of defining pairs up to 20. The 137 F_{hk0} were used to construct the xy projection of the electron density. The signs of all the F_{hk0} were calculated from the coordinates given by this projection; in seven cases the signs were reversed. The new xy projection (from all 158 amplitudes) gave the

Fig. 7. The xy electron-density projection for gadolinite.

final x and y coordinates. The divergence coefficient (neglecting zeros) was 17.9%. The zeros up to sin ϑ/λ = 1.1 raised R to 22.4%.

We use the h0l reflection to find the z coordinates. The Weissenberg photograph (on b) for the zero-layer line gave 154 nonzero reflections, of which 22 had $|U_{h0l}|$ exceeding 0.5. The a plane again enabled us to halve the h. The h0l reflections complied with $S_{H \pm K} = S_H \cdot S_K$ in the sets

	5̄03	800×303		5̄09	500×009
(301̄)	307	004×303	(8̄09)	809	800×009
(204̄)	4010	004×303	(6̄07)	6.0.11	602×009
	707	404×303	(7̄07)	907	800×107
(301̄)	907	604×303	(604)	4̄010	5̄03×107
(6̄02)	1204	901×303	(705)	5̄09	6̄02×107
(306)	3012	009×303	(703)	5̄011	6̄04×107
(901)	9̄05	602×303	(602)	0.0.12	3̄05×307
(11.0.1̄)	507	8̄04×303	(3.0.11)	303	004̄×307
(11.0.3̄)	509	806×303		12.0.4	800×404
	2̄08	505×303	(1006)	2̄02	6̄02×404
(10.1.1)	505̄	208×303		903	507×404
	602̄	305×303	(1.0.10)	11.0.4	503̄×607
(2̄.0.10)	804̄	507×303	(11.0.1)	107	503̄×604
(2̄06)	800	503×303	(008)	12.0.0	604̄×604
	1200	903×303	(12.0.0)	004	602×6̄02
	6011	607×004	(103)	1.0.11	107×004
(9̄01)	907	903×004	(507̄)	707	107×600
(1̄2.0.4)	12.0.4	12.0.0×004	(303̄)	903	303×600
(501)	5̄07	5̄03×004	(307̄)	907	307×600
	5̄09	505×004		11.0.5̄	505×600
(503̄)	5.0.11	507×004		0.0.12	009×003
	8̄04	800×004	(1̄09)	1̄105	507̄×602
(905)	9̄03	901×004	(301)	905	303̄×602̄
	6̄02	602×004	(600)	006	303̄×303
(3.0.11)	3̄03	307×004		406	103×303
(6̄06)	602	602̄×004		0.0.12	309̄×303
	901	903̄×004		6011	308×303
	6.0.11	602×009	(12.0.2̄)	6̄08	905̄×303
	2̄.0.11	2̄02×009	(6̄04)	12.0.2	901̄×303
(607)	6̄.0.11	6̄02×009		10.0.6	10.0.2×004

Then we found that all the reflections appearing in the inequalities could be expressed in terms of three letters. Amplitudes with h = 2m, k = 2n were either plus or b. The signs of amplitudes with h = 2m, k = 2n + 1 were c and bc. Also, those with h = 2m + 1, k = 2n had signs of ac or abc. If both indices were odd, the signs were a and ab.

We used the known and checked signs for the hk0 reflections (for which R was small) to find the absolute sign of b (=−), i.e., that sign which does not depend on the choice of origin, since this letter defines the sign of 16.0.0 ($U_{16.0.0}$), which is known to be minus from the xy projection. Thus, the signs of the amplitudes with h = 2m, k = 2n were found to be both plus and minus, while the other amplitudes were defined by a, c and ac. We checked that a = c = + was correct from 10.0.0, which is common to the xy and xz projections (its sign in xy was minus, and here was abc with b = −1, so a = c); this gave us a reference group of 79 reflections. Zachariasen's method gave a further 66 signs. The xz electron-density projection was constructed from 145 reflections; here the B peak was sharper than was the Be peak for herderite, and was also displaced further towards the base of its tetrahedron (relative to Be), which agrees with the tendency of B to compromise between fourfold and threefold coordination [19]. The signs of three F_{h0l} were changed in the recalculation. A repeat Fourier synthesis (with 154 amplitudes) gave a more regular relation between the peak heights; some of the peaks were displaced slightly. R for the 154 non-zero h0l was 12.9%; or 16.5% with the 100 zero ones, again up to sin ϑ/λ = 1.1.

Table 1 gives the atomic coordinates derived from the xy and xz projections and also gives Ito and Mori's results, which differ from ours by up to 0.025 (0.12 in the z coordinates). The two kinds of tetrahedron are much

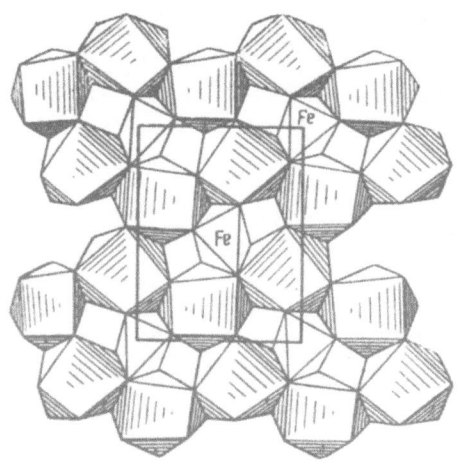

Fig. 8. The xy projection for gadolinite in Pauling's polyhedra. The lower layer has the Y polyhedra; the Fe octahedra are at the centers of symmetry 000 and $\frac{1}{2}$ $\frac{1}{2}$ 0.

more regular; the distances are $Si - O = 1.57$, 1.58, 1.63, 1.63 A; all three $B - O = 1.48$ A (as against 1.45, 1.44, 1.47, in Ito's paper); $Ca - O$ falls within the limits 2.35-2.66 A. The different $O - OH$ distances (2.35, 2.52) given by Ito are replaced by 2.46 A. $B - OH = 1.54$ A.

The structure of the beryllophosphate (herderite) is related to that of the borosilicate (datolite) in such a way that the replacement of $Be^{2+}P^{5+}$ by $B^{3+}Si^{4+}$ corresponds to a change from the smaller PO_4 tetrahedron (edge about 2.50A) to the larger SiO_4 one (edge about 2.61 A), which change is balanced by the change from the large BeO_3F tetrahedron (edge about 2.64 A) to the smaller $BO_3(OH)$ one (edge about 2.45 A).

The same direct methods were used to solve for the structure of the not entirely isostructural rare-earth mineral gadolinite. The formula is derived from that of datolite, $CaBOHSiO_4$, by replacing $Ca^{2+}B^{3+}$ by $Y^{3+}Be^{2+}$ and by inserting at the centers of symmetry in the unit cell (which contains four "datolite" molecules) two extra $Fe^{\cdot\cdot}$ ions (in Fig. 7 the $Fe^{\cdot\cdot}$ octahedra are shown with the Y polyhedra), which insertion is compensated by replacing four OH by O. The resulting formula is $Fe^{\cdot\cdot}Y_2Be_2O_2Si_2O_8$ with two such molecules in the cell ($a = 9.89$, $b = 7.52$, $c = 4.71$ A; $\beta = 90°33'$). Ito gives no atomic coordinates; he demonstrates the structural similarities only by comparing in a general way the strengths of ten Debye lines and certain other lines.

Gadolinite is appreciably radioactive, which was checked on our samples; the cause is that thorium replaces the rare earths isomorphously, which results in the mineral's becoming metamict (i.e., to a large extent semi-amorphous), and which made the x-ray patterns of very poor quality. It is possible to analyze the structure, but the intensities could be estimated only for the hk0 reflections (111 nonzero), which gave 16 with $|U_{hk0}|$ larger than 0.5. Taking those initial $|U_{hk0}|$ larger than 0.40, we had 28 inequalities, which were strong ones, and which gave a group of 40 reference signs. The statistical treatment then left only 12 signs undefined. The resulting electron-density pattern (Fig. 7) was very clear. Table 1 gives the x and y coordinates. These are fairly accurate ($R = 18.2\%$, neglecting the zero reflections, or 22.9%, taking account of them up to $\sin\vartheta/\lambda = 0.8$) and may be used to examine the differences between gadolinite and datolite, which arise mainly from the extra $Fe^{\cdot\cdot}$ ions at the empty centers of symmetry in the Ca polyhedra in the first layer (Fig. 8). The Ca has also been replaced by the smaller Y, and the B by the larger Be, of course. Table 1 shows that the two structures differ very much in the coordinates of O_I, O_{II}, O_{IV}, Si and Y. The $Fe - O$ distances calculated for gadolinite from the data for datolite are $Fe^{\cdot\cdot} - O = 1.79$, $Fe^{\cdot\cdot} - O_{IV} = 2.03$ and $Fe^{\cdot\cdot} - O_V = 1.74$ A. Except for $Fe^{\cdot\cdot} - O_{IV} = 2.03$ A, the distances are clearly impossible for silicates (the sum of the ionic radii of Fe^{2+} and O^{2-} is $0.80 + 1.32 = 2.12$ A), and so the z coordinates must differ considerably between the two structures, the more so since the replacement of Ca^{2+} by $\overline{Y}^{3+}$ ($r_Y = 0.97$ A, $r_{Ca} = 1.04$ A) must reduce the $Y - O$ distance of gadolinite relative to the $Ca - O$ distance of datolite, whereas the $Fe^{\cdot\cdot}$ ions at the centers of the four-rings must push the nearby O atoms sufficiently to give acceptable $Fe^{\cdot\cdot} - O$ distances. This appears as the displacement of O_{II} and O_{IV} from the center, which is visible even in the xy projection. The interatomic distances may be calculated from the coordinates taken from the projection of gadolinite, but with our z coordinates taken from datolite, which at once gives more reasonable distances, namely, $Fe^{\cdot\cdot} - O_{II} = 1.92$, $Fe^{\cdot\cdot} - O_{IV} = 1.98$ and $Fe^{\cdot\cdot} - O_V = 2.18$ A.

In the structures of herderite and datolite we find grids with strong bonds (P, Si, B) alternating with ones with weak bonds (Ca polyhedra with large distances between cation and anion), whereas in gadolinite the second grids are stronger because Ca^{2+} replaces Y^{3+}, and also because Fe (which absorbs light strongly) is present. The result is that gadolinite is much harder and is optically positive, whereas herderite and datolite are optically negative.

LITERATURE CITED

[1] A. G. Betekhtin, Mineralogy [in Russian] (Moscow, 1950).

[2] E. S. Dana, Descriptive Mineralogy [Russian translation] (ONTI, 1937).

[3] Strukturbericht IV, 169, 203.

[4] H. Strunz, Mineralogische Tabellen (Leipzig, 1949).

[5] T. Ito and H. Mori, Acta Cryst. 6, 24 (1953).

[6] M. M. Umanskii and S. S. Kvitka, Izv. AN SSSR, Ser. Fiz. 15, 147 (1951).

[7] Kh. S. Mamedov and V. Belov, Zap. Vses. Min. Obshch. 85, 1 (1956).

[8] N. V. Belov and I. M. Rumanova, Trudy Inst. Krist. 9, 103 (1945).

[9] G. S. Zhdanov and V. P. Kotov, Zhur. Fiz. Khim. 15, 918 (1941).

[10] D. Harker and J. Kasper, Acta Cryst. 1, 70 (1948).

[11] A. J. C. Wilson, Nature 150, 151 (1942).

[12] B. K. Vainshtein, Structural Electron Diffraction [in Russian] (Izd. AN SSSR, 1956).

[13] B. K. Vainshtein, J. Exptl.-Theoret. Phys. (USSR) 27, 44 (1954).

[14] H. Lipson and W. Cochran, The Determination of Crystal Structures [Russian translation] (IL, 1956).

[15] S. V. Borisov, P. V. Pavlov and N. V. Belov, Kristallografiya 3, 1, 90 (1958).*

[16] W. Zachariasen, Acta Cryst. 5, 68 (1952).

[17] E. Schiebold, "The structure of silicates," Coll.: Basic Concepts of Geochemistry.3 [Russian translation] (Leningrad, 1937).

[18] N. V. Belov, The Structures of Ionic Crystals and of Metallic Phases [in Russian] (Izd. AN SSSR, 1947).

[19] V. S. Sobolev, Introduction to the Mineralogy of Silicates [in Russian] (Lvov, 1949).

---

* Soviet Physics — Crystallography, p. 85.

Reprinted from Doklady of the Academy of Sciences of the USSR (Doklady Akademii Nauk SSSR),
Earth Sciences Sections, Vol. 130, pp. 167-170, January-February, 1960

CRYSTAL STRUCTURE OF LOVENITE*

V. I. Simonov and Academician N. V. Belov

Crystallographic Institute of the Academy of Sciences of the USSR

In the well-known reference books [1, 2], lovenite is placed in the group of "nezo" (ortho) silicates with complex compositions with a general formula of the type $A_3B[SiO_4]_2$ (O, F, OH), where A = Ca, Na and B = Zr, Ti, Nb. The samples investigated by us — crystals of lovenite from the Lovozersk massif with an increased Ti content (titanolovenite) — were obtained from Ye. I. Semenov, who, in accordance with the analysis of M. Ye. Kazakova, gave its detailed formula as [3]

$$Na_{1.23}Ca_{0.94}Mn_{0.33}Fe^{2+}_{0.20}Fe^{3+}_{0.11}Ti_{0.26}$$

$$Nb_{0.09}Zr_{0.73}[SiO_4]_2(F, OH).$$

Laue diagrams and Weissenberg photographs of the layer lines along the b and c axes (Mo radiation) confirm that lovenite is monoclinic with the holohedral Fedorov space group $C^5_{2h} = P2_1/a$, which is simply determined by the regular absences of reflections. The parameters a = 10.54°A, b = 9.90A, c = 7.14A, β = 108°12' agree with previous data [1, 2] if account is taken of the considerably increased quantity of Ti in titanolovenite. The unit cell contains n ≈ 4 formula units of the composition discussed. Despite the orthosilicate character of the gross formula, we suggested earlier that all the minerals of the group lovenite — velerite to the Ca-diorthosilicate, cuspidine, $Ca_4Si_2O_7F_2$, [6], were similar in structure.

The complete interpretation of the lovenite structure was carried out by the superposition of Patterson projections. Analysis of p(x, y) and p(x, z) ruled out the possibility of placing the cations at the centers of symmetry and so three other atoms are linked with each basic atom in (x, y, z) symmetry; and these groups of four atoms fix important Patterson peaks with coordinates and weights: 1) $(2x, 2y, 2z) - Z^2$; 2) $(2x, 2\bar{y}, 2z) - Z^2$; 3) $(\frac{1}{2}, \frac{1}{2} + 2y, 0)$; 4) $(\frac{1}{2} + 2x, \frac{1}{2}, 2z) - 2Z^2$. The peaks with weight Z^2 correspond to the vectors between atoms connected to the center of symmetry and these are very useful as starting points for the superposition method [7]. In searching for these peaks in the Patterson diagram their elementary ratio is useful: the sum and difference of vectors 3 and 4 (~ twice as large by weight) give the necessary vectors. The key to the structure of lovenite is the projection along the short c axis, i.e., p(x,y); in it were found two peaks of types 1 and 2 and minimalization was carried out on them. The presence of elements of glide symmetry permitted the construction of

two functions $M_4(x, y)$ as the result of minimalization and further, by combination of the latter, $M_8(x, y)$ was obtained from which were got the atomic coordinates for the initial planar model of the structure.

In this projection the diorthogroup Si_2O_7 is revealed as the axis parallel to the direction of the projection (c = 7.14A) and it is very exactly overlapped by octahedra of cations formed from the continuous ribbons of tetrahedra which also stretch along the c axis and resemble the ribbons in cuspidine. The precision was increased by the usual method of successive projections of electron density. From the known x and y coordinates of all atoms it did not prove difficult to look at the Patterson chart p(x, z) as the starting point for minimalization of peaks. The starting point for increasing the precision of the z coordinates was determined with the $M_4(x, z)$ function. The final projections of the electron density of lovenite are shown in Fig. 1.

The structure of lovenite is yet another illustration of the concept of the "second chapter in the Crystal Chemistry of the Silicates" [8], in which the determining role is played by the equality in size of the Na and Ca octahedra and the height of the diorthogroups Si_2O_7. As in cuspidine [6], the basic structural features of lovenite are the four-fold continuous ribbons constructed from cation octahedra.

Their positions with respect to one another are shown in Fig. 2, in which for greater clarity the oxygen octahedra in the columns are placed exactly one above the other, whereas the coordinates of the overlapping O atoms are somewhat different (Table 1). In the hollow tube between the ribbons are placed diorthosilicate groups in pairs which are moved along the height c in such a way that they contact the edges of three of the octahedra around the large cations.

The structure was determined with 45 parameters of 60 atoms in the cell. Apart from 2Si and 9 anions, 4 cations are introduced as basic atoms. In cuspidine, 4 Ca atoms occupy these positions, whereas in lovenite we have:

$$Na_{1.23}Ca_{0.94}Mn_{0.33}Fe^{2+}_{0.20}Fe^{3+}_{0.11}Ti_{0.26}Nb_{0.09}Zr_{0.73}.$$

The question of the distribution of these cations among the 4 crystallographically independent positions is of

---

* Translation of: Kristallicheskaya Struktura Lovenita, Doklady Akademii Nauk SSSR, vol. 130, no. 6, 1960, pp. 1333-1336.

TABLE 1. Coordinates of the Basic Atoms of Lovenite (in fractions of the axes x, y, z)

Atom	x	y	z	Atom	x	y	z
Zr	0.294	0.105	0.024	O_{III}	0,250	0,241	0.800
(Fe, Mn)	0.237	0.376	0.856	O_{IV}	0.163	0.483	0.180
Ca	0.304	0.105	0.525	O_V	0.122	0.483	0.750
Na	0.425	0.376	0.343	O_{VI}	0.480	0.223	0.098
Si_I	0.123	0.331	0.223	O_{VII}	0.484	0.239	0,670
Si_{II}	0.117	0.331	0.667	O_{VIII}	0.119	0.010	0,945
O_I	0.130	0.331	0.443	(F, OH)	0.120	0.005	0,423
O_{II}	0.238	0.241	0.197				

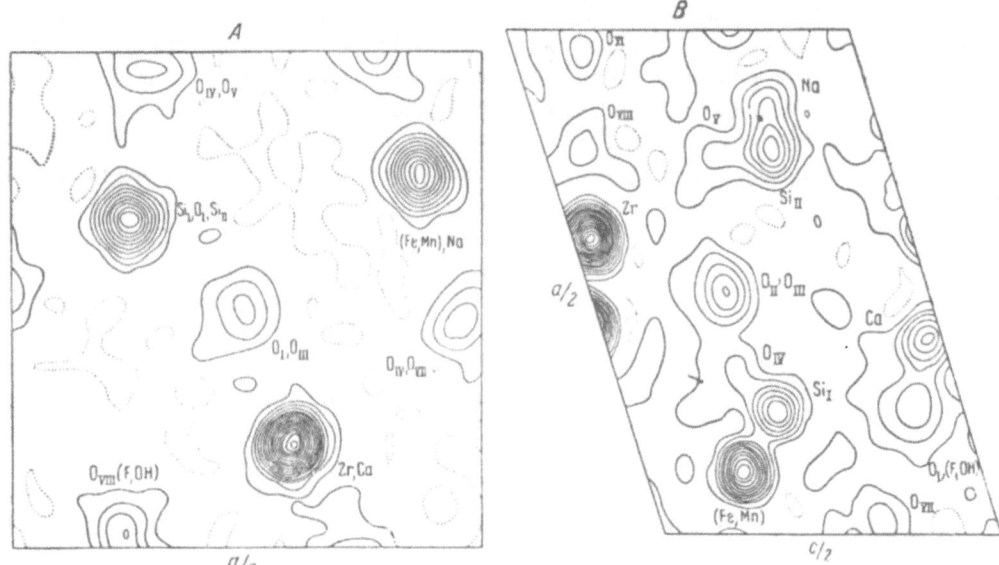

Fig. 1. Final projection of the electron density of lovenite. Isolines constructed through 8 el/A². A — σ(x, y), B — σ(x, z).

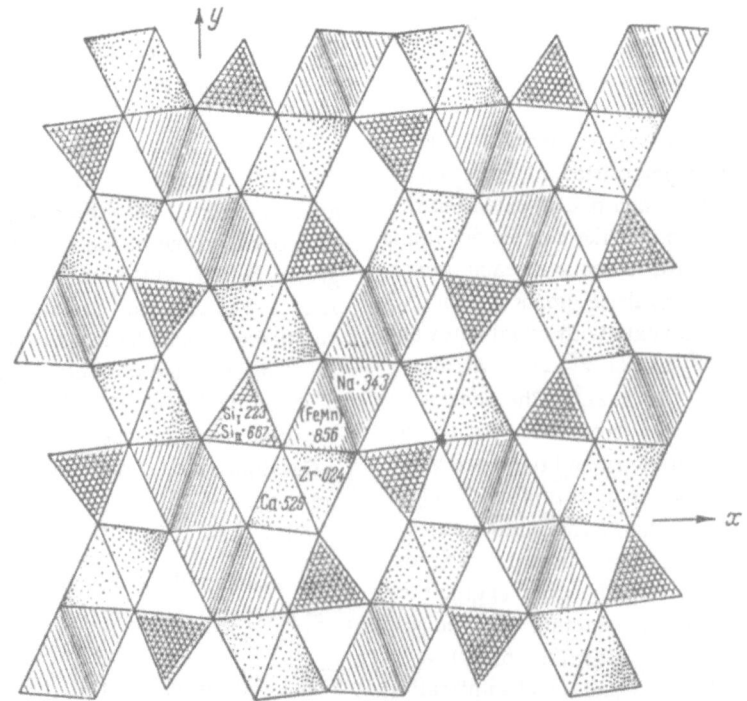

Fig. 2. Plan of the lovenite structure in polyhedra.

interest in itself. The ratio of the heights of the maxima on $\sigma(x, y)$ and $\sigma(x, z)$ permits the assertion that $Zr(Z = 40)$ is concentrated at one position; two are taken by Na and Ca (and in fact one of these positions has only Na) and all the rest of the cations are accumulated at the fourth position. Zr in lovenite occupies a fourth of the positions, but in the cell there are only ~ 3 atoms of Zr. As has been shown in the interpretation of the structure of seidozerite, the isomorphous replacement of Zr by titanium takes place in the presence of Mn[9]. If this is applied to the chemical composition of titano-lovenite from one deposit [3],

$$Na_{1.34}Ca_{0.75}Mn_{0.57}Fe^{2+}_{0.27}Ti_{0.55}Nb_{0.09}Zr_{0.58}[SiO_4]_2$$
$$(F, OH),$$

then comparison indeed shows the agreement in the changed quantity of Ti and Zr with the correspondingly increased amount of Mn. If it is supposed that in our case $(Zr_{0.73}, Ti_{0.26})$ occupies one place then 0.25 Na or Ca must be introduced into $(Mn_{0.33}, Fe_{0.31}, Nb_{0.09})$. In another variant of the cation distribution which is in agreement with the peak heights in the projection, Ca is added to the Zr cations, whereas Ti is amalgamated with Mn and Fe. The atomic numbers of Ti and Ca are low (Z = 22 and 20), and since the problem concerns their addition to Zr(Z = 40), the x-ray method using visual estimation of the intensity does not permit a definite choice to be made between the two possibilities. By calculation of all (148) nonzero structure amplitudes the divergence of the $h0l$ factor equals 20.5% for the first and 20.0% for the second variant. The $\sigma(x, y)$ projection is still less suitable for the solution of the problem because of the almost exact pairing of the overlapping cations, although its divergence factor equals

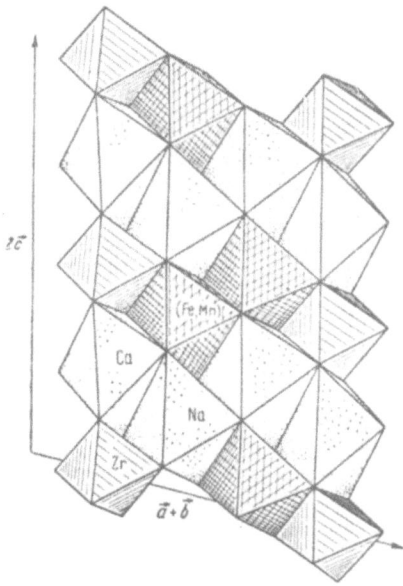

Fig. 3. A cation ribbon to show the position of the various cations in the octahedra.

12.1% (nonzero hk0 171). Thus, the admission of visually estimated intensities only with two layer lines does not allow the question to be decided unequivocally and therefore in Figs. 1-3 and Table 1 each occupied position is marked with the symbol of the atom which plays the basic part in the creation of Z_{eff} of the corresponding "average atom."

In Fig. 3 is shown the pattern of occupation with the different cation of the octahedra in the ribbons. In each of the four continuous columns the ribbons of octahedra make contact with common (horizontal) edges. In the outer columns of the ribbon, Zr and Ca octahedra alternate and in the inner columns (Fe, Mn) and Na octahedra, the (FeMn) octahedra of adjoining columns being joined in pairs by a common edge. Similar colums of octahedra are characteristic of a series of structures determined in recent years. Large octahedra usually alternate with average-sized ones along the column and the corresponding repetition period of two octahedra is 7.14A in lovenite, 7.10A in seidozerite, and 7.53A in cuspidine where both octahedra are occupied by the large Ca.

The Si-O distances in the diorthogroup are within the limits 1.55-1.67A; in the Zr octahedra, 5 Zr-O distances are 1.99-2.05A, the sixth is 2.19A; in the (Fe, Mn) octahedra: 2.07-2.31A; all 12 cation-anion distances in the Na and Ca octahedra are within the limits 2.10-2.62A.

Starting from the crystal structure of lovenite and inclining to the first of the two ways of distributing the cations described above, we propose that its detailed chemical formula should be written in the form $Na_{0.97} \cdot (Ca_{0.70}Na_{0.26})(Fe_{0.31}Nb_{0.09}Mn_{0.33}Ca_{0.24})(Zr_{0.73}Ti_{0.26}) \cdot O(F, OH)[Si_2O_7]$ with islands (nezosilicate) of Si_2O_7. The eighth O is not in the siloxy radical, i.e., it plays a role similar to that of the ninth anion (F, OH). On recalculating the chemical analysis of the starting material [3] there were two atoms of Si and the insignificant deficit of cations could be explained by excess of SiO_2 having been omitted. The accumulation of ~ 24% of the large cation Ca with the Fe, Nb, and Mn distorts the structure and this is certainly reflected in the constants of the temperature factor which were obtained in the final stage by comparison of the experimental and calculated intensities. For the projection $\sigma(x, y)$ along the axis, B is equal to $0 \cdot 87A^2$ for both the diorthogroup and the columns of octahedra, whereas for $\sigma(x, z)$, B = $= 1.35 A^2$. It is readily suggested that the very marked increase of the constant in the second case originates from the addition of the statistically disordered distribution of the atoms (with substantially different radii) along the column of octahedra to the thermal oscillations.

REFERENCES

1. Betekhtin, A. G., Mineralogiya, Moscow, 1950, p. 719.

2. Strunz, H., Mineralogische Tabellen, 1957, S. 273.

3. Semenov, Ye. I., M. Ye. Kazakova, and V. I. Simonov, Zap. Vsesoyuzn. min. obskch, vol. 87, no. 5, 1958, p. 540.

4. Simonov, V. I., and N. V. Belov, Kristallografiya vol. 4, no. 2, 1959, p. 162.

5. Mamedov, Kh. S., V. I. Simonov, and N. V. Belov, Doklady Akademii Nauk SSSR vol. 126, no. 2, 1959, p. 379.

6. Smirniva, P. F., I. M. Rumanova, and N. V. Belov, Zap. Vsesoyuzn. min. obshch., vol. 84, no. 2, 1955, p. 159.

7. Simonov, V. I., Kristallografiya, vol. 4, no. 3, 1959, p. 302.

8. Belov, N. V., Min. sborn. Lvovsk. geol. obshch., no. 13, 1959, p. 23.

9. Belov, N. V. and V. I. Simonov, Doklady Akademii Nauk SSSR, vol. 125, no. 4, 1959, p. 888.

Received, December 7, 1959

Reprinted from Soviet Physics—Crystallography, Vol. 5, pp. 186-198, September-October, 1960

DERIVATION OF THE STRUCTURE OF LOVOZERITE
FROM SECTIONS OF THE THREE-DIMENSIONAL
PATTERSON FUNCTION

V. V. Ilyukhin and N. V. Belov

Institute of Crystallography, Academy of Sciences
Translated from Kristallografiya, Vol. 5, No. 2, pp. 200-214, March-April, 1960
Original article submitted January 6, 1960

The structure of lovozerite, $Na_2ZrSi_6O_{15} \cdot 3H_2O \cdot 0.5\ NaOH$, has been deduced from sections of the electron-density pattern. The silicon-oxygen radical is shown to be a $[Si_6(O, OH)_{18}]$, ring; the silicon-oxygen tetrahedron contains five or six OH groups.

Lovozerite, an extremely rare mineral, was first found in 1935 as grains associated with eudialite in the Lovozero alkali complex (central Kola peninsula) by Gerasimovskii, who described it in 1939 [1].

Specimens much altered by weathering have a black color and fracture irregularly. The specific gravity is 2.394 (pycnometer), the hardness is about 5, and cleavage is not observed. Powder patterns taken with Fe radiation showed no lines; on the basis of optical data (negative uniaxial with $n_\gamma = 1.560$, $n_\alpha = 1.549$) it has been assigned to the hexagonal system [1]. Gerasimovskii proposed an empirical formula

$$[H, Na, K]_2O \cdot (Ca, Mg, Mn)\ O \cdot (Zr, Ti)\ O_2 \cdot 6SiO_2 \cdot 5H_2O \rightarrow Na_2CaZrSi_6O_{16} \cdot 5H_2O,$$

from analyses of grains deriving from an occurrence on the River Muruai; that formula is the one given in the literature [2].

In composition lovozerite is similar to other Zr silicates of the eudialite-catapleiite group, especially to eudialite $Na_{4-5}CaFeZr(OH,Cl)Si_6O_{18}$, eucolite (eudialite containing much divalent metal and Nb), and mesodialyte [3] (which is intermediate in composition); its distinguishing features are a larger amount of water and a smaller amount of alkali. Studies on material found later in other localities here (by Tikhonenkov, in the Kola peninsula) and abroad (Greenland) have given similar results. Semenov [4] has reported single-crystal fragments of brown-red color occurring as overgrowths on eudialite. These latter crystals were in a much better state of preservation; they gave a more exact specific gravity (2.64) and a powder pattern (Fe radiation). The optical constants were confirmed, and a new analysis was obtained (analyst I. Razina), which showed a much higher content of alkali (12.4% as against 5.6-6%), more silica (55.6% instead of 52%), and less water (8.23% against 13-15%). Rare-earths were also detected.

The most probable composition is

$$Na_2ZrSi_6O_{15} \cdot 3H_2O$$

with additional NaOH (about half a molecule).

This formula is close to that of elpidite ($Na_2ZrSi_6O_{15} \cdot 3H_2O$) and to that of dalyite ($K_2ZrSi_6O_{15}$), which contains no water; these two representatives of the eudialite group differ from the other Zr silicates in containing either Na alone or K alone (within the usual limits of purity for any mineral).

Fig. 1. The p(xz) Patterson projection.
The contours are in arbitrary units.
Negative contours have been omitted.

Fig. 2. Projection of the Patterson function on the
xy plane. The contours are in arbitrary units.
Negative contours have been omitted.

I

The irregular single-crystals received from E. I. Semenov included three of roughly spherical shape (0.3-0.8 mm in size), which gave good Laue patterns as well as oscillation and rotation photographs; from these the class was found to be C_{2h} = 2/m, the cell parameters being a = 10.48, b = 10.20, c = 7.33 A, β = 92°30' with two (1.96) formula units in the unit cell.

The Weissenberg photographs (Mo radiation) showed absences only for h + k = 2n + 1, which corresponds to the x-ray group 2/mC − / −; that group includes one space group having a center of symmetry and two (Cm, C2) lacking such center.

The intensities were estimated on the usual blackening scale modified to have close intervals of $\sqrt[4]{2}$. Those spots falling at small ϑ/λ were concealed by the clamp in the goniometer cassette; these were estimated separately from patterns recorded with Cu radiation. As the crystals were nearly spherical, the corrections for absorption must be nearly the same for all the main zones, so they were neglected.

The 116 nonzero h0l reflections, 117 similar hk0, and 78 similar 0kl gave us three basic Patterson projections, but only the xz one showed a set of three peaks [5] characteristic of a centrosymmetric projection. Statistical analysis of the experimental structure factors [6] confirmed that the xz distribution is centrosymmetric and that the other two are not. There is one heavy atom in the independent part of the cell [7]; allowance for that did not alter the results deduced from [6]. The intensity distribution calculated from all the hkl also lacked a center of symmetry, while the one-dimensional Harker trace taken along y through the three-dimensional pattern showed none of the peaks it ought to if there were an m plane normal to b.

So, although the crystal shows no piezoelectric effect, it would appear that there is no m plane, the most probable space group being C_2^3 = C2, with screw and rotation axes alternating. The independent part of the cell contains only two kinds of twofold positions having each a single parameter (on crystallographically different twofold rotation axes); the other positions are general and have three parameters. The chemical formula would imply that one of the one-parameter positions must be occupied by Zr and the other by Na, the other atoms being in general positions.

Figures 1 and 2 show the xz and xy projections respectively. The primitive cell contains one heavy atom (Z_{Zr} = 40); the restrictions placed on the symmetry of the group are such as to cause us to hope that the structure might be deducible directly from the projections, because the Zr-X vector system gives peaks much stronger than those given by the Na-X, Si-X, etc. systems. But unfortunately the pattern of peaks is regular on account of the special configuration of the structure; peaks are superimposed because atoms overlap (the axes of the cell are large), and many peaks are broadened on account of the nearness to a centrosymmetric array; in consequence we can draw no reliable conclusion about the positions of the atoms.

The xz projection (b = 10.2 A) has its peaks arranged in bands parallel to the coordinate axes. The sole peak in p(xz) suitable for use in the superposition method [8] is the Zr-Si one ($\underline{x}$ in Fig. 1); but even $M_2(xz)$ retains all the peaks in p(xz)[$M_1(xz)$ does as well, of course], so no conclusion can be drawn. This result is confirmed by Simonov's [9] test for the possibility of locating the atoms:

$$Z_x Z_{max} > 4b \sqrt{|\bar{P}|^2}, \quad \text{where} \, |\bar{P}|^2 = \frac{1}{S^2} \sum F_{h0l}^4$$

(with allowance for the zeroth term). We refer the $|F_{h0l}|$ to the absolute scale and estimate $|\bar{P}|^2$; we get

$$Z_x Z_{max} > 3880 \, e^2/A^2.$$

Now the strongest Zr-Si peak has [23] a strength of 550 e^2/A^2; four peaks are superimposed in the synthesis (on account of the resemblance to a centered structure), but still 4 × 550 < $Z_x Z_{max}$.

The higher symmetry of the xy and xz projections (the traces of two symmetry planes are present) does not make them convenient for minimalization. At first sight the xy projection appears suitable (c = 7.33 A along the axis), but it lacks single peaks and has very pronounced diagonal bands, so it also fails to give a sharp $M_n(xy)$. Again, the yz projection fails because two structures are superimposed in the M function; those structures are generated by that function, which is centrosymmetric.

In view of the presence of a horizontal mirror plane in the vector space we replaced the usual projections

$$\int_0^b P(xyz) \cos 2\pi ky/b \, dy \quad \text{and} \quad \int_0^b P(xyz) \sin 2\pi ky/b \, dy$$

by modified ones [10]

$$\int_0^{b/2} P(xyz) \cos 2\pi ky/b \, dy \quad \text{and} \quad \int_0^{b/2} P(xyz) \sin 2\pi ky/b \, dy$$

(and by the sum and difference of those), for which purpose we used h0l and h2l reflections. The peaks were broad and clearly overlapped, so we could not interpret them correctly. The statistical method [11, 12] also failed.

In view of these successive failures we were forced to use P(xyz), the three-dimensional Patterson function. For that purpose we had results for 13 planes normal to the axes in the reciprocal lattice, namely seven layer lines for $\underline{b}$, five for $\underline{c}$, and one (equatorial) for $\underline{a}$. In all there were 1850 reflections corrected for angle factors and referred to a single scale.

The three-dimensional function has been applied with success for centrosymmetric structures [13-15], but for structures lacking centers (space groups P2_1, P$2_12_12_1$, P$6_1$2) use has been made only of Harker sections [16-18] or of sections of the function showing peaks caused by pairs of atoms related by symmetry elements [19].

The peaks in the xy and xz projections were such as to indicate that suitable sections would be ones taken normal to the twofold axis at heights (in hundredths of $\underline{b}$) of 0, 8, 10, 12, 13, 15, 17, 20, 22, 23, 25, 27, 28, 30, 33, 35, 37, 38, 40, 42, 50.[*]

Figure 3 shows the independent parts of the more important sections (those containing the strongest peaks); Fig. 4 shows all the main vectors from Zr by means of a dimetric projection of a sphere [20]. The size of a spot indicates the strength of the peak; the number indicates the length of the vector; and the orientation may be

[*]We wish to thank Professor G. S. Zhdanov and his colleagues in the computing division of the Karpov Institute of Physical Chemistry for permission to use their 'Kristall' computer.

Fig. 3. Sections of P(xyz), the three-dimensional Patterson function.
a) y = 0; b) y = 0.13; c) y = 0.22; d) y = 0.25.

read in ρ, φ coordinates from the equiangular grid. The two unique Zr atoms in the side-centered cell are related by an oblique translation; one may be considered as being at the origin 000 (the other is at $\frac{1}{2}\frac{1}{2}$ 0), so the Zr-X peak gives the absolute coordinates of the X atom. The sign of r_x (the orientation to be chosen from the two possible ones) is to be established from considerations of chemical crystallography and so on (i.e., by trial and error).

The assignment of any peak was checked by calculating (by means of integral characteristics [21]) the absolute height of that peak at its center,

$$P_{ij}(0) = \frac{1}{2\pi^2} \int_0^\infty f_i(s) f_j(s) s^2 \, ds,$$

and the shape of the peak

$$P_{ij}(r) = \frac{1}{2\pi^2} \int_0^\infty f_i(s) f_j(s) s^2 \frac{\sin sr}{sr} \, ds \approx \frac{1}{2\pi^2} \sum_{k=1}^{K_{\max}} f_i(s_k) f_j(s_k) s_k^2 \frac{\sin s_k r}{s_k r} \Delta s_k;$$

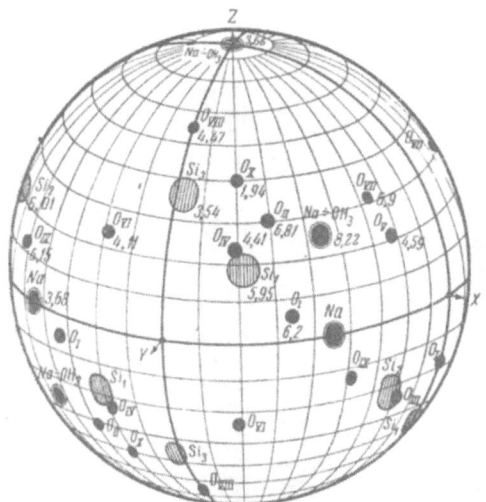

Fig. 4. The main Zr-X interatomic vectors.

Fig. 5. Electron-density synthesis σ(xz). The contours are at intervals of 7 e/A², or of 3.5 e/A² for the partly overlapping O atoms.

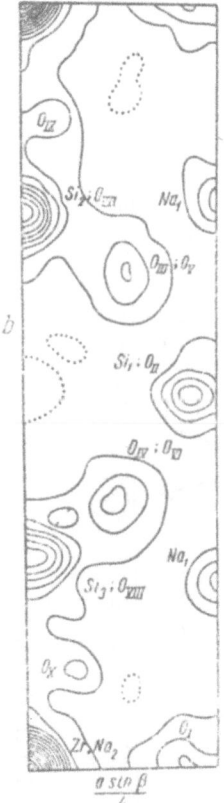

Fig. 6. Electron-density synthesis σ(xy). The contours are at intervals of 7e/A².

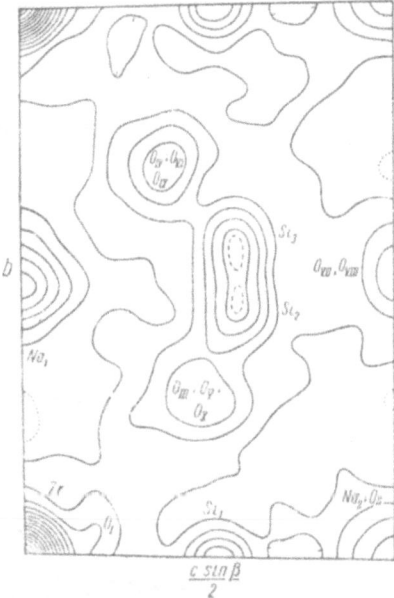

Fig. 7. Electron-density synthesis σ(yz). The contours are at intervals of 7 e/A², or 3.5 e/A² for adjacent Si atoms.

this process was applied to the Zr-Si, Zr-Na, Zr-O, Si-Si, Si-O, O-O peaks subject to a certain amount of averaging (over h0l , hk$l$, hk0, 0kl zones and the whole of the space of the reciprocal lattice hk$l$) for the temperature factor B computed by means of Wilson's method [22]. For this purpose we took the mean isotropic temperature factor for all atoms to be B = 1.75 A².

The peaks on P(xyz) were referred to the absolute scale. For that purpose we used a slightly modified formula from [23]:

$$KP(000)_0 = \sum_{i=1}^{N} P_{ii}(0) - \frac{F_{000}^2}{V} \, ,$$

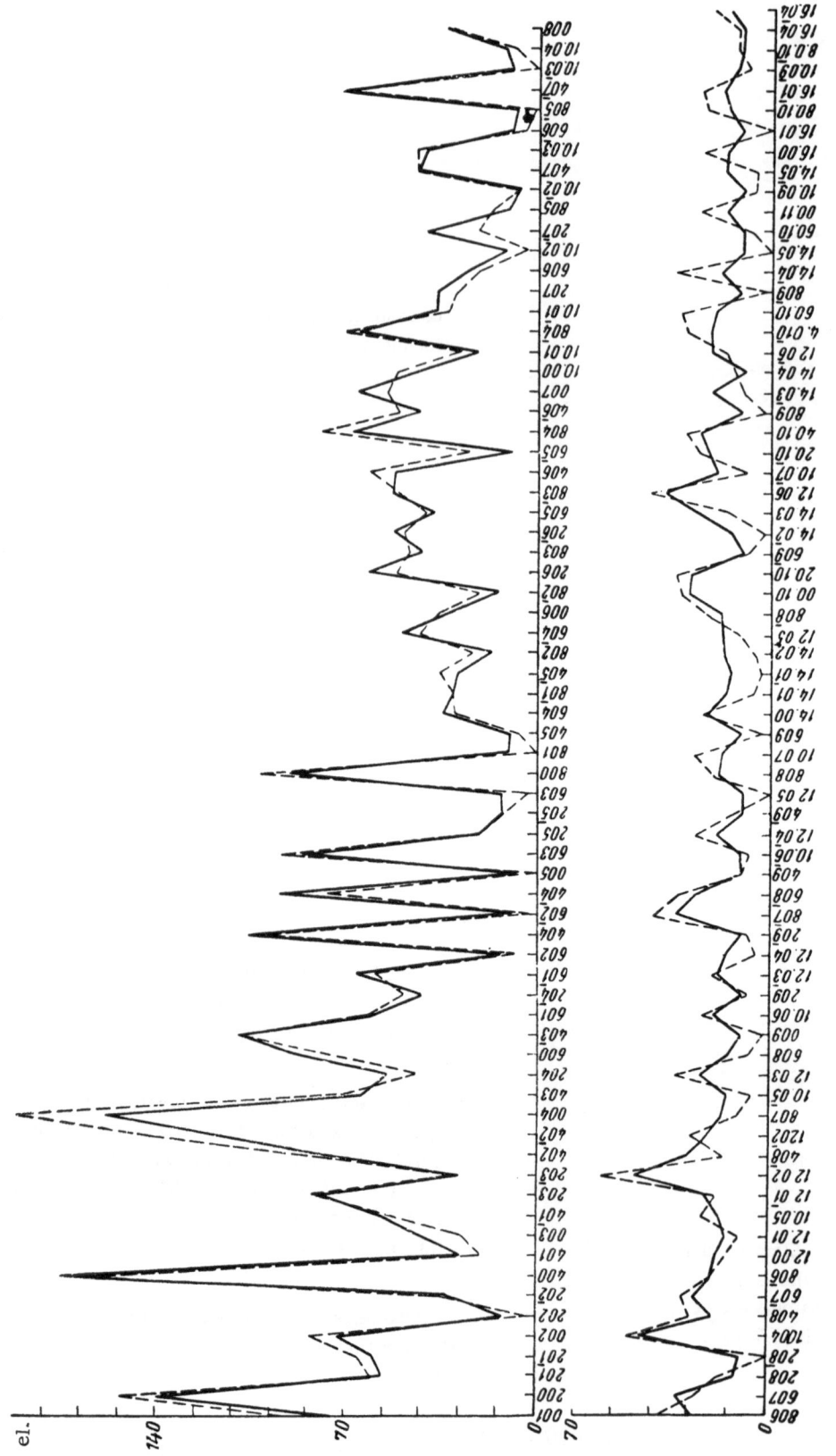

Fig. 8. Structure amplitudes for the h0*l* zone.

—— F_0 - - - - F_c.

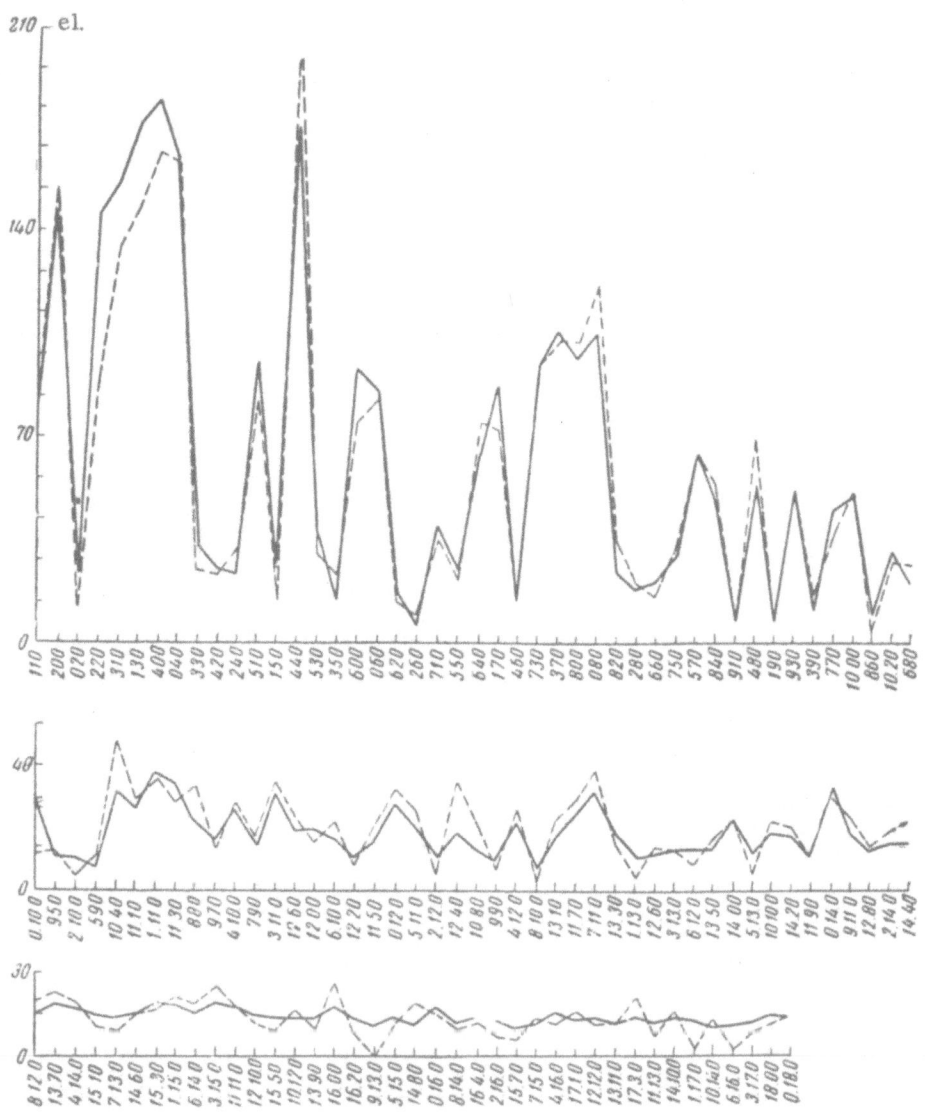

Fig. 9. Structure amplitudes for the hk0 zone. —— F_0 ---- F.

TABLE 1

Atom	x^*	y^*	z^*	Atom	x^*	y^*	z^*
Zr	0	0	0	O_{VI}	61.0	84.0	81.3
Si_1	21.1	49.5	26.5	O_{VII}	50.0	27.5	50.0
Si_2	49.5	23.2	70.9	O_{VIII}	50.0	74.5	50.0
Si_3	48	78.2	70.9	O_{IX}	4.0	85.5	17.6
O_I	18.8	2.0	89.3	O_X	6.6	13.5	19.6
O_{II}	23.0	49.5	53.0	Na_1'	24.8	24.6	00.0
O_{III}	37.1	13.5	74.0	Na_1''	25.2	74.6	00.0
O_{IV}	37.5	87.0	78 0	$(OH_3) Na_2$	00.0	1.5	50.0
O_V	62.8	17.0	75.4				

*Coordinates given in hundredths of <u>a</u>, <u>b</u>, and <u>c</u>.

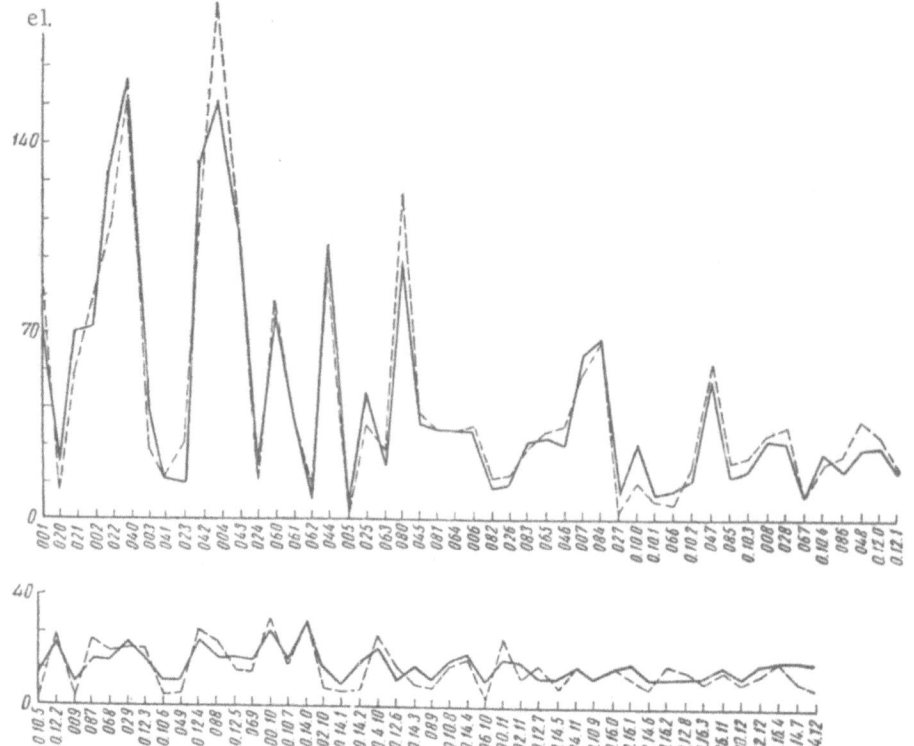

el.

Fig. 10. Structure amplitudes for the 0k*l* zone.

—— F₀ ---- F.

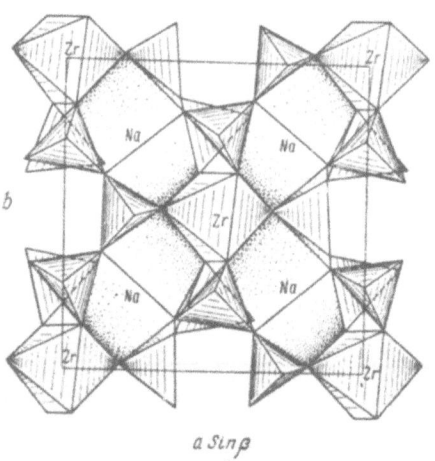

Fig. 11. Projection of the structure of lovozerite (in polyhedra) on the xy plane.

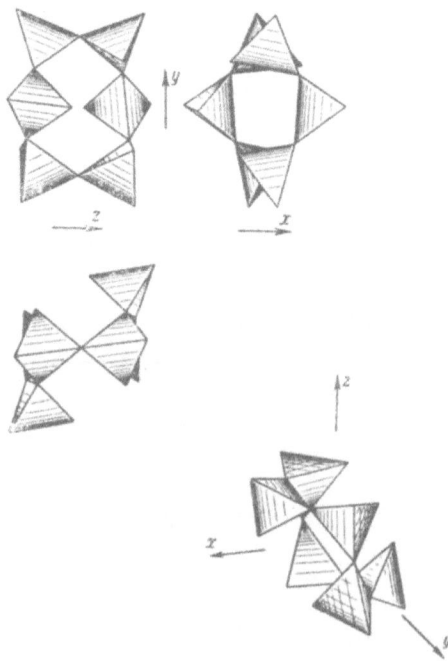

Fig. 12. Ring of silicon-oxygen tetrahedra (projections on the xz, xy, and yz planes, and axonometric projection).

TABLE 2

	Neglecting zero reflections (%)	With zero reflections (%)	No. of zeros
R_{h0l}	18.0	20.2	24
R_{hk0}	18.5	19.5	12
R_{0kl}	18.0	19.4	15

in which $P(000)_b$ is the value of the function (unnormalized) at the origin of the Patterson cell, $F_{000}^2 = \left(\sum_{i=1}^{N} Z_i \right)^2$,

V is the volume of the cell, and $P_{ii}(0)$ is calculated from certain formulas [21, 21]. Then the absolute value of $P(xyz)$ is

$$P(x, y, z)_{abs} = K P(xyz)_b + \frac{F_{000}^2}{V}$$

The extra term $\dfrac{F_{000}^2}{V}$ appears because the zeroth term cannot be incorporated in $P(xyz)_b$ since it cannot be given in relative units.

The peaks agreed well in height and shape with those found from experiment, so we were able to assign preliminary coordinates to Si, Na, and some of the oxygen atoms; in that way we obtained a first rough model.

TABLE 3

Anion	Cation							$\Sigma \frac{w_i}{n_i}$
	Zr	Na₁	Na₁	Si₁	Si₂	Si₃	Na₂	
O_I	$4/6$	$1/6\,(1/7)$		$4/4$				$2 - \frac{1}{6}\left(\frac{1}{7}\right)$
O_{II}				$4/4$			$1/6$	$1 + \frac{1}{6}$
O_{III}		$1/6\,(1/7)$		$4/4$	$4/4$			$2 + \frac{1}{6}\left(\frac{1}{7}\right)$
O_{IV}			$1/6\,(1/7)$	$4/4$		$4/4$		$2 + \frac{1}{6}\left(\frac{1}{7}\right)$
O_V		$1/6\,(1/7)$			$4/4$			$1 + \frac{1}{6}\left(\frac{1}{7}\right)$
O_{VI}			$1/6\,(1/7)$					$1 + \frac{1}{6}\left(\frac{1}{7}\right)$
$O_{VII-VIII}$					$4/4$		$1/6$	$2 + \frac{1}{6}$
O_{IX}	$4/6$	$1/6\,(1/7)$			$4/4$		$1/6$	$2\left(2 - \frac{1}{42}\right)$
O_X	$4/6$		$1/6\,(1/7)$			$4/4$	$1/6$	$2\left(2 - \frac{1}{42}\right)$

We calculated a series of electron-density syntheses for the xz (centrosymmetric) projection*, for which purpose a was divided into 120 parts and c into 60 parts; then we proceeded to the xy and yz projections (in the first we divided b and a into 60 parts; in the second, b into 120 parts and c into 60 parts). Six successive syntheses were performed for $\sigma(x, z)$, four for $\sigma(y, z)$, and two for $\sigma(x, y)$ (Figs. 5-7); as a result we established the coordinates of all the main atoms and were able to allow for displacement arising from overlapping [24]. Table 1 gives the coordinates resulting from syntheses carried to the points at which R attained the values given in Table 2.

*The work was done at first manually and later on the 'Strela' universal computer in the Computing Center at Moscow State University; we are deeply indebted to the team at the Center (especially V. L. Simonov) for their assistance.

TABLE 4

Zr Octahedron (A)

$Zr - O_I = 2.16$ $\quad O_I - O_{IX} = 3.15$
$\qquad\quad O_{IX} = 2.05$ $\qquad\quad - O_X = 2.87$
$\qquad\quad O_X = 2.13$ $\qquad\quad - O_{IX}^* = 3.00$
$\qquad\qquad\qquad\qquad - O_X^* = 3.00$
$\qquad\qquad\quad O_{IX} - O_{IX}^* = 3.26$
$\qquad\qquad\quad O_X - O_X^* = 2.79$
$\qquad\qquad\quad O_{IX} - O_X = 2.88$

Si_1 Tetrahedron (A)

$Si_1 - O_I = 1.62$ $\quad O_I - O_{II} = 2.72$
$\qquad\quad O_{II} = 1.64$ $\qquad\quad - O_{III} = 2.55$
$\qquad\quad O_{III} = 1.67$ $\qquad\quad - O_{IV} = 2.66$
$\qquad\quad O_{IV} = 1.61$ $\quad O_{II} - O_{III} = 2.60$
$\qquad\qquad\qquad\qquad - O_{IV} = 2.74$
$\qquad\qquad\quad O_{III} - O_{IV} = 2.73$

Si_2 Tetrahedron (A)

$Si_2 - O_{III} = 1.65$ $\quad O_{III} - O_V = 2.72$
$\qquad\quad O_V = 1.58$ $\qquad\quad - O_{VII} = 2.68$
$\qquad\quad O_{VII} = 1.60$ $\qquad\quad - O_{IX}^* = 2.53$
$\qquad\quad O_{IX} = 1.57$ $\quad O_V - O_{VII} = 2.58$
$\qquad\qquad\qquad\qquad - O_{IX}^* = 2.63$
$\qquad\qquad\quad O_{VII} - O_{IX}^* = 2.56$

Si_3 Tetrahedron (A)

$Si_3 - O_{IV} = 1.54$ $\quad O_{IV} - O_{VI} = 2.50$
$\qquad\quad O_{VI} = 1.69$ $\qquad\quad - O_{VIII} = 2.77$
$\qquad\quad O_{VIII} = 1.60$ $\qquad\quad - O_X^* = 2.52$
$\qquad\quad O_X = 1.72$ $\quad O_{VI} - O_{VIII} = 2.77$
$\qquad\qquad\qquad\qquad - O_X^* = 2.78$
$\qquad\qquad\quad O_{VIII} - O_X^* = 2.57$

Na_1 Polyhedron (A)

$Na_1 - O_I = 2.52$ $\quad O_I - O_{III} = 2.55$ $\quad O_V - O_{IX}^* = 3.76$
$\qquad\quad O_{III} = 2.60$ $\qquad\quad - O_V = 3.64$ $\qquad\quad - O_X = 3.25$
$\qquad\quad O_{IV} = 2.46$ $\qquad\quad - O_X = 2.87$ $\qquad\quad - O_I^* = 3.78$
$\qquad\quad O_V = 2.40$ $\quad O_{III} - O_V = 3.72$ $\quad O_{VI}^* - O_{IX}^* = 3.70$
$\qquad\quad O_{VI} = 2.25$ $\qquad\quad - O_{VI}^* = 3.51$ $\qquad\quad - O_X = 3.54$
$\qquad\quad O_{IX} = 2.66$ $\qquad\quad - O_{IX}^* = 2.52$ $\qquad\quad - O_I^* = 3.60$
$\qquad\quad O_X = 2.68$ $\quad O_{IV} - O_{VI}^* = 2.91$ $\quad O_{IX} - O_I^* = 3.14$
$\qquad\quad O_I = 3.00$ $\qquad\quad - O_X = 2.50$ $\quad O_{IV}^* - O_V = 3.30$
$\qquad\qquad\qquad\qquad - O_I^* = 2.66$

Na_2 — OH_3 Polyhedron (A)

$Na_2 - O_{II} = 2.44$ $\quad O_{II} - O_{VII} = 3.32$ $\quad O_{IX} - O_{VII}^* = 2.55$
$\qquad\quad O_{VII} = 2.46$ $\qquad\quad - O_{VIII}^* = 3.53$ $\qquad\quad - O_X = 2.88$
$\qquad\quad O_{VIII} = 2.36$ $\qquad\quad - O_{IX} = 3.64$ $\quad O_{VIII}^* - O_X = 2.60$
$\qquad\quad O_X = 2.65$ $\qquad\quad - O_X = 3.37$
$\qquad\quad O_{IX} = 2.93$ $\qquad\quad O_{IX}^* = 3.09$
$\qquad\qquad\qquad\qquad O_X^* = 4.03$

Note. The stars denote atoms other than the basic ones, to which they are related by the symmetry elements, however.

These values are roughly equal for the three independent projections on account of the method of estimating the intensities[**] and of the isometric nature of the structure, which is responsible for the identical absorption (see description of structure).

[**] At first we tried to use the xy and yz projections, supposing that the derivative from central symmetry would fail to make itself felt. But five successive revisions of the coordinates would not reduce R below 45%. We obtained satisfactory values of R only when the symmetry was reduced to C2.

The coordinates are determined by 41 parameters. The three ρ (r), projections were supplemented from the weak peaks in P(xyz) in order to revise the coordinates of overlapping atoms.

The error [23] varies somewhat from one projection to another:

$h0l$, $0kl$: Zr $\pm$ 0.0013 A; Si $\pm$ 0.005 A; Na $\pm$ 0.0065 A; O $\pm$ 0.0012 A
$hk0$: Zr $\pm$ 0.001 A; Si $\pm$ 0.004 A; Na $\pm$ 0.0055 A; O $\pm$ 0.01 A.

Figures 8-10 give the experimental structure amplitudes calculated from the coordinates given in Table 1 and referred to the absolute scale.

The structure of the first Zr silicate of the eudialite group (catapleiite, $NaZr[Si_3O_9]2H_2O$) was established (in 1936) by B. K. Brunovskii in this laboratory [25]; its symmetry $D^4_{6h} = P6_3/mcm$. It consists of almost flat layers of cations joined by a hexagonal array of Zr octahedra and Na decahedra; the layers are separated by gaps of 5 A, which is about half $\underline{c}$. Between the layers there are triple (single-layer) rings of $[Si_3O_9]$tetrahedra. Two H_2O molecules are distributed randomly over three possible positions (2/3 in each).

Our description of the structure of the second mineral in this group (lovozerite) must start with the xy projection (Fig. 11). It is clear that Zr octahedra and Na polyhedra alternate along the [110] diagonals of the cell to form an almost exactly tetragonal array having a square of side 7.31 A. In the xy projections two such identical squares are related by an oblique translation of group C2. The Zr atom lies at the center of a slightly distorted oxygen octahedron (at the corner of a square); the Na atom (at the middle of a side) is surrounded by a characteristic figure having seven vertices, which is made up a trigonal prism and half an octahedron joined on a common square face [compare Ca (Na) in sphene, durangite, epidote, etc.] The eighth (somewhat more remote) anion gives a figure composed of two distorted trigonal prisms joined at their bases. That figure is very convenient to use in representing the structure (Fig. 8), although in fact each rectangular face is bent along its diagonals. The flat grids are separated vertically by c sin β, i.e., along the axis c = 7.33 A but with a displacement through the angle β. At half $\underline{c}$, between Zr atoms, lies another Na atom (that appearing in the second part of the formula); but chemical analysis gives a half molecule of NaOH in the 'side chain', and the structure factors indicate that this Na atom is replaced in half the instances by OH or (more probably) by OH_3.*

The main feature of lovozerite is the three-dimensional lattice of Zr octahedra lying at the corners of cubic cells 7.31 A $\times$ 7.31 A $\times$ 7.33 A having a slight monoclinic inclination relative to the base along a diagonal (β is 92°30', not 90°). At the middle of each edge there is an Na atom, which along c are replaced in half the cases by OH or OH_3. This isometric framework is responsible for the lack of cleavage.

A silicon-oxygen radical of closed type lies within each (almost) cubic cell. It is a six-member ring of silica units $[Si_6O_{18}]$ (not a three-member ring, as in catapleiite) and is of the type found in beryl and dioptase. The related Zr silicate elpidite has been supposed (but not shown) to contain a two-level trigonal ring (doubled catapleiite ring). This six-member ring may also be considered as being two catapleiite rings (the doubling causes the distance between the cation layers to rise to 1.5 $\times$ 5 $\approx$ 7.33 A), but the rings are not closed, the two being related by a horizontal twofold axis on [010] (Fig. 12). This six-member ring may also be represented as two Si_2O_7 groups parallel to $\underline{c}$ and closed only by two Si tetrahedra (one above, one below). The pseudo-sixfold axis of the ring lies along [$\bar{2}$01]. The two such rings in the monoclinic cell are parallel and are related by the oblique translation in that cell.

The mineral is to be reckoned as unstable on account of this structure of the silicon-oxygen radical, in conjunction with the mobile Na polyhedra and heavy Zr nodes (which together give the structure little strength); it is readily destroyed by weathering. A useful measure of this lack of strength is its constant B = 1.75, which is very high for a silicate; this value shows that the thermal oscillations are large even at room temperature (other Zr silicates have B $\leq$ 1) and that the structure is open and of intermediate type.

The radical has the formula $[Si_6O_{18}]$ found also for beryl and dioptase; in each of the six tetrahedra two oxygen corners are common to adjacent tetrahedra (the other two are connected to one Si atom only); but the rings in the two latter minerals are of high symmetry (6/m and 6° = $\bar{3}$), whereas the ring in lovozerite has only

*This 50% substitution of OH or OH_3 for Na causes R to fall by 1.5%.

a single twofold axis lying in its own plane. Each Si tetrahedron has only one of its O corners entering a Zr octahedron, so the formula for the mineral and the balance of the valences imply that one corner must be taken by an OH group; in all the ring contains six hydroxyl groups, which appear in the tetrahedra in very much the way for afwillite [26], pectolite [27], $Ca_2[SiO_3OH]OH$ [28], and epididymite [29]. The proper formula for the radical is $[Si_6O_{12}(OH)_6]$, which makes lovozerite the fifth silicate in which Si is joined directly to hydroxyl. The number of groups in the radical may be reduced to five (randomly distributed over the six positions) if we assume that the groups replacing 50% of the Na atoms on the vertical edges are OH_3, rather than OH.

The R are very good, and so is the balance of the valences, in view of the unusual structure of the radical.

We saw above that the Na lie in figures with eight corners, but one O is further away from Na_1, and two from Na_2; it would be better to say that $\frac{1}{7}$ of the valence strength of Na_1 goes to each O($\frac{1}{8}$ for Na_2). For simplicity in the calculations we have taken the figure to be $\frac{1}{8}$ in both cases; in that case we get for each oxygen a sum differing from two (or one) by much less than the 25% usually taken to be the limit permissible under Pauling's second rule. The deviation from balance is even less in every case (except for O_{IX} and O_X, for which it becomes $1/_{42}$) if we use $1/_7$ instead of $1/_6$. Three oxygen atoms (O_{II}, O_V, O_{VI}) are bonded to one Si only (are not bonded to Zr); the sum over the cations for each of them is about one, so they must be in OH groups.

Table 4 gives all the interatomic distances, which agree well with those found for other silicates.

The Zr octahedron has on its twofold axis three Zr-O distances, which are 2.05, 2.13, 2.16 A; now Zr^{4+} + O^{2-} = 2.15 A, so the edges of the octahedron are 2.79-3.26 A long.

The slight scatter about the means found for the Si tetrahedra (Si-O = 1.54-1.72 A and O-O = 2.50-2.78 A) are to be explained as caused by hydroxyl, even though the Si-OH distance is nearly equal to the Si-O one, as in afwillite and pectolite.

The two Na cations in the main part of the formula have seven neighbors at distances of 2.25-2.68 A and have eight at 3 A. Six Na-O distances for the additional Na are 2.34, 2.44, 2.65 A, with two further Na-O of 2.93 A.

An interesting point is that a slight deformation of (almost) cubic cell (to make it triclinic), if such as to join the free corners of the tetrahedra in one ring with similar corners in another, would make a chain of tetrahedra corresponding to the radical $[Si_6O_{15}]$ (here we have a possible model for dalyite $K_2ZrSi_6O_{15}$).

The closeness of the structure to cubic (especially to centrosymmetric) affects the F^2 syntheses very greatly; we have pointed out above how greatly the banding etc., hinder us in deducing the structure.

Lovozerite gives us another example of a mineral in which the directions of the principal refractive indices coincide with the diagonals of the x-ray cell (not with the axes). The identical array of Zr and Na atoms found along both diagonals of the (001) faces is responsible for making two refractive indices equal (n_γ = 1.560), the third being lower (on account of the weakening of the edges of the cube along c); the result is that the mineral was at first considered to be uniaxial, the symmetry being over-estimated as hexagonal or tetragonal.

LITERATURE CITED

[1] V. I. Gerasimovskii, Doklady Akad. Nauk SSSR 25, 751 (1939).

[2] A. G. Betekhtin, Mineralogy [in Russian] (Moscow, 1950).

[3] M. D. Dorfman, Dissertation [in Russian] (1959).

[4] E. I. Semenov, Trudy Kol'sk. Fil. Akad. Nauk SSSR (1960).

[5] Kh. S. Mamedov and N. V. Belov, Doklady Akad. Nauk SSSR 106, No. 3 (1956).

[6] E. R. Howells et al., Acta Cryst. 3, 210 (1950).

[7] G. A. Sim, Acta Cryst. 11, 123 (1958).

[8] M. J. Buerger, Acta Cryst. 4, 531 (1951).

[9] V. I. Simonov, Kristallografiya 4, No. 3 (1959).*

*[See Soviet Physics - Crystallography, vol. 4].

[10] I. M. Rumanova, Kristallografiya $\underline{3}$, No. 6 (1958).*

[11] W. H. Zachariasen, Acta Cryst. $\underline{5}$, 60 (1952).

[12] E. G. Fesenko, I. M. Rumanova, and N V. Belov, Kristallografiya $\underline{1}$, No. 2 (1956). **

[13] K. Robinson, Acta Cryst. $\underline{7}$, 494 (1954).

[14] P. G. Taylor and C. A. Beevers, Acta Cryst. $\underline{5}$, 341 (1952).

[15] J. A. Ibers, Acta Cryst. $\underline{9}$, 225 (1956).

[16] L. Nitta, K. Sakurai, and Y. Tomiie, Acta. Cryst. $\underline{4}$, 289 (1951).

[17] A. E. Smith, Acta Cryst. $\underline{5}$, 224 (1952).

[18] S. v. Houten and E. H. Wiebenga, Acta Cryst. $\underline{10}$, 156 (1957).

[19] D. Harker, J. Chem. Phys. $\underline{4}$, 381 (1936).

[20] E. A. Glazunov and N. F. Chetverukhin, Axonometry [in Russian] (Moscow, 1953).

[21] V. V. Ilyukhin and S. V. Borisov, Strukt. Khim. $\underline{1}$, No. 1 (1960).

[22] A. J. C. Wilson, Nature $\underline{150}$, 152 (1942).

[23] B. K. Vainshtein, Zhur. Éksp i Teor. Fiz. $\underline{27}$, 44 (1954).

[24] A. D. Booth, Proc. Roy. Soc. A, $\underline{188}$, 77 (1946).

[25] B. K. Brunovskii, Acta Physicochem. USSR $\underline{5}$, 863 (1936).

[26] H. D. Megaw, Acta Cryst. $\underline{5}$, 477 (1952).

[27] M. J. Buerger, Z. Krist. $\underline{108}$, 248 (1956).

[28] L. Heller, Acta Cryst. $\underline{5}$, 725 (1952).

[29] E. A. Pobedimskaya and N. V. Belov, Doklady Akad. Nauk SSSR $\underline{129}$, No. 4 (1959).

*[See Soviet Physics- Crystallography, vol. 3].
**[See Soviet Physics- Crystallography, vol. 1].

Reprinted from Journal of Structural Chemistry, Vol. 1, pp. 44-54, May-June, 1960

THE STRUCTURE OF EPIDIDYMITE NaBeSi$_3$O$_7$ (OH).
A NEW TYPE OF [Si$_6$O$_{15}$] CHAIN

E.A. Pobedimskaya and N.V. Belov

Lomonosov State University, Moscow
Translated from Zhurnal Strukturnoi Khimii, Vol. 1, No. 1, pp. 51-63,
May-June, 1960
Original article submitted December 25, 1959

An x-ray study has been made of epididymite NaBeSi$_3$O$_7$(OH). Ordinary and weighted projections have been used to deduce the structure. The coordinates of all Si, Na, O, and Be atoms have been found. The structure is based on columns of Na polyhedra having strips of composition Si$_6$O$_{15}$ running parallel; these strips are duplicated wollastonite chains. The Si$_6$O$_{15}$ strips combine with Be tetrahedra to form double layers perpendicular to c.

Epididymite is a rare Na-Be silicate which has been found in the USSR in the Kola peninsula [1]. Recent studies agree with Gossner and Kraus's finding [2] that epididymite (and especially the monoclinic form thereof, eudidymite) resembles the feldspars very considerably (in density and optics), but there is also a considerable resemblance to the micas [3], which appears as perfect cleavage on (001), tabular habit, and pseudohexagonal behavior.

Ito [4] proposed a structure in 1934; the main feature was the [Si$_3$O$_8$]$_\infty$ chains (Fig. 1) running parallel to b = 7.34 A*. This was only a tentative proposal, made before Patterson methods were available; the pattern was in agreement with the cell parameters and with the symmetry, but there was no proper x-ray confirmation. The proposed structure had many defects; the Si-Si distances in the silicon-oxygen tetrahedra varied from 1.11 to 2.80 A, and other unacceptable features were Be-Si and Be-O separations of 0.54 and 0.72 A respectively. The Na atom located at a center of symmetry was enclosed (on the common faces of the Na octahedron) by two Si tetrahedra (Fig. 2). Wyckoff [8] pointed out that the structure must be revised.

We had available specimens from Yu. A. Pyatenko's collection and also specimens we collected ourselves during 1958 near Lovozero (Kola peninsula). We used single crystals 0.3-0.5 mm in size, all of which were in the form of plates or tablets. The main results have been derived from eight Weissenberg photographs on b and two (the zeroth and first) on a. Good results were not obtained on c; the spots were irregular in shape, and there were many spots in excess [both effects are caused by the resemblance to mica; the layers are displaced along (001), the perfect cleavage direction].

---

*Berman [5] proposed a chain of this type for minerals of the wollastonite group (for xonotlite in particular); b is about 7.3 A in nearly every case. Mamedov and Belov rejected Berman's model on the basis of their results for xonotlite [6] and wollastonite itself [7]; they proposed a metasilicate chain [SiO$_3$]$_\infty$=[Si$_{2+1}$O$_9$]$_\infty$ for wollastonite, this chain differing only in the length of its links from the chain found in diopside. Nevertheless, the geometry of the silicon-oxygen radical in epididymite is closely related to the corresponding feature in wollastonite (see below).

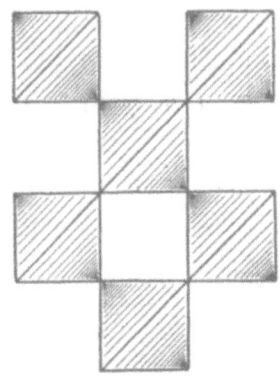

Fig. 1. Berman's $[Si_3O_8]_\infty$ chain.

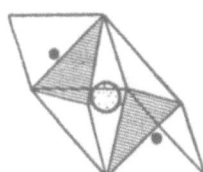

Fig. 2. Combination of a Na octahedron with two Si tetrahedra in Ito's structure for epididymite.

The cell parameters are a = 12.66, b = 7.34, c = 13.48 A, which are close to Ito's values [4]; they agree well with the axial ratios given in 1894 by Flink [9] if we double c. The specific gravity (2.55) implies that the cell contains eight formula units $HNaBeSi_3O_8$.

The space group is D_{2h}^{16} = Pnam, as Zachariasen claimed in 1929 [10]. Tests for piezoelectric response (made in the Department of Crystal Physics at Moscow State University) gave negative results, which pointed to the presence of a center of symmetry, as did the intensity statistics [11] and the results obtained by the triplets method applied to the Patterson projections [12].

The resemblance to mica [10] (c is a pseudosixfold axis) appears in the ratio a:b (which is almost $\sqrt{3}$) and in the weakness of the reflections with h+k = 2n+1 in the xy projection. The latter feature indicates that the lattice is pseudocentered on the C (= ab) plane, as we would expect.

The intensities were estimated visually by reference to 35 calibrated spots. We used different radiations (Mo and Cu) to determine the intensities for the zero layer lines, and then calculated mean values. The strong reflections were corrected by reference to the β spots. The structure amplitudes were calculated in the usual way with allowance for the polarization and kinematic factors. The $|F|^2_{max}$: $|F|^2_{min}$ were in the ratio 1000: 1, approximately. Absorption was neglected in calculating the $|F|^2$, because this mineral is effectively semitransparent to x rays and because the specimens were roughly isometric.

The spots numbered about 2000, and the atoms are all similar in atomic number, so one would expect that Zachariasen's statistical methods would solve the structure. But this laborious work did not give the expected result, because the atoms lie almost entirely in layers (see [13]) and the mineral is pseudohexagonal. Harker-Kasper methods could not be used, because the reference group contained only four reflections having absolute amplitudes ≥0.5. We were forced to use only Patterson functions.

The short axis is b, so we would expect that the P(uw) projection would show the fewest overlapping peaks; but Fig. 3* shows that the pattern is too complex to be interpreted without ambiguity, so we constructed weighted projecting P_1 (uw) and P_2 (uw) (from 303 reflections of h1l type and from 226 of h2l type respectively) in order to ease the task of locating the first-order** Si-Si peaks in P(uw). These projections had only one system of first-order peaks; they gave the x and z of the "heaviest" atom (Si) uniquely. The results were confirmed by taking Harker sections at the levels 0 and ½ along b, for which purpose we used 846 reflections from the first, second, and third layer lines appearing on the rotation photographs taken on b.

We constructed P(vw) (Fig. 4) from 68 0kl reflections and P_1 (vw) from 202 1kl reflections in order to find the third coodinate of this Si. Next we constructed M(uw), the minimal function [15], by superposition. Minimalization of P(uw) gave the coordinates of all the Si, and of six O out of ten. The M(vw) gave only the coordinates of the silicon and those less reliably, because the peaks were less sharp in the vw projection.

The Patterson diagram gave only the coordinates of Si_I unambiguously, so we used these most reliable coordinates to calculate the signs, without reference to the results from the minimalization, in order to construct the electron-density projection along b. Figure 5 shows the projection on the xz plane; we see that Si_I and Si_{III} coincide in σ(xz). The peak for Si_{III} has the same z as the first two, but x is doubled: $x_{I,II} = 10/60$; $x_{III} = 20/60$. This projection, in addition, shows peaks corresponding to the O atoms; almost all peaks in M(uw) have corresponding peaks in σ(xz). There is a rather large peak at the origin in σ(xz), which we at first

---

* The a' is halved in this projection as a result of the a glide plane normal to b.

** Porai-Koshits's name [14] for the peaks produced by the vectors between atoms in the same regular system.

Fig. 3. Epididymite. Patterson projection along <u>b</u>.

Fig. 4. Patterson projection along <u>a</u>.

took to represent Na; but the volume available for the Na polyhedron is inadequate, so the Na cannot lie at the inversion center. We found later that this peak corresponds to two Be atoms superimposed.

The electron-density projection constructed with allowance for the coordinates of three Si, ten O, and one Be had a further peak, which we naturally took to be Na. The $\underline{x}$ and $\underline{z}$ were revised, and then we calculated the structure amplitudes, which are compared with the experimental ones in Fig. 6. The superimposed pairs for Si (I and III) and O appearing in the xz projection facilitated the work on the projection along $\underline{b}$, but it hindered the process of refining the coordinates, so R for the h0l reflections remained fairly large: 0.23 (B = 0.56 A^2, $\sin \theta/\lambda \leq 1.1$); when 17 zero-intensity reflections were allowed for ($\sin \theta/\lambda \leq 0.9$) we found that R was 0.24.

Three Si atoms were located by minimizing P (vw); the approximate $\underline{y}$ and the reliable $\underline{z}$ were used to calculate the signs for the 0kl zone. The electron-density projection on $\underline{a}$ (Fig. 7) gave revised $\underline{y}$ and confirmed most of the $\underline{z}$. An error-series synthesis was used to refine the coordinates of Be and O. Here R for the projection along $\underline{a}$ was 0.21 ($\sin \theta/\lambda \leq 1$) or 0.225 with allowance for 12 zero-intensity reflections. Figure 8 gives the calculated and experimental 0kl structure amplitudes.

Table 1 gives the coordinates of the 15 basal-plane atoms. Except for the three kinds of O atoms (which lie in mirror planes), all atoms (including Na, of which there is only one kind - see above) lie in general eight-fold positions, so the total number of parameters is $(15 \times 3) - (3 \times 1) = 42$.

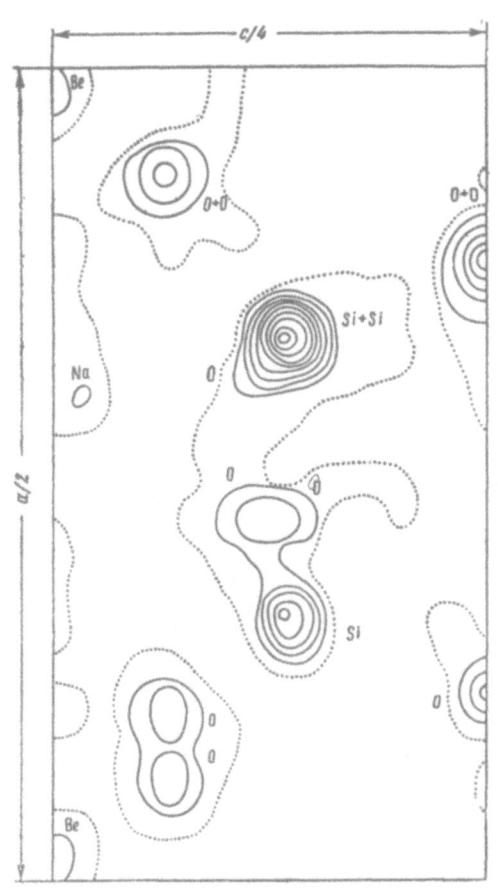

Fig. 5. Electron-density projection along b.

The Si-O distances vary from 1.59 to 1.74 A, and the Na-O from 2.1 to 2.7 A (compare the 2.14 to 2.62 A of [13]). The probable errors are given by Vainshtein's formula [16] as Si $\pm$ 0.004 A, Na $\pm$ 0.006 A, O $\pm$ 0.009 A, Be $\pm$ 0.023 A.

The Si-O-Si angles are 130; 134; 135.5; 139; 149 and 160°.

Table 2 gives the distances in each coordination polyhedron. Table 3 gives the same for Ito's coordinates. Table 4 gives the balance of the valencies for each of the ten kinds of O(OH) atoms (see Table 1).

The mineral resembles most other silicates [17] in having columns of oxygen polyhydra arranged around a large cation (here Na). Here the polyhedra have eight corners and take the form of the twisted cubes so characteristic of Ca grossular garnet and of the Na-Li analog cryolithionite (CuAl$_2$ [17] is composed exclusively of such polyhedra). The columns of Na polyhedra extend along the twofold screw axes parallel to b. Two such polyhedra occur in one b repeat distance (Fig. 9a), as in many silicates having columns of Na-Ca polyhedra that have been examined recently*; all of these structures have b very close to 7.3 A.

Here b = 7.3 A, and at three levels along that axis there are three SiO$_4$ tetrahedra, which form a link of a wollastonite chain [Si$_3$O$_9$] = [Si$_{2+1}$O$_9$] in which two tetrahedra face one way and one the other.

Two such chains are joined into a band** as in xonotlite Ca$_6$[Si$_6$O$_{17}$] (OH)$_2$ [6], although only every third tetrahedron is linked in the latter case, because the symmetry plane is normal to the plane containing the centers of all three tetrahedra in a link, whereas the symmetry plane in epididymite is parallel to the plane containing the centers, so all three tetrahedra in a link are joined. If xonotlite has 2 [Si$_3$O$_9$] − O = [Si$_6$O$_{17}$] for the formula of its

TABLE 1

Relative Coordinates of the Basal Atoms in Epididymite†

Atom	x	y	z	Atom	x	y	z
Si$_I$	0,167	0,500	0,150	IV	0,075	0,017	0,075
Si$_{II}$	0,334	0,207	0,142	V	0,117	0,900	0,250
Si$_{III}$	0,152	0,877	0,132	VI	0,075	0,410	0,075
Na	0,194	0,167	0,492	VII	0,278	0,400	0,150
Be	0,025	0,215	0,032	VIII	0,200	0,684	0,095
I	0,400	0,225	0,042	IX	0,267	0,010	0,125
II	0,117	0,483	0,250	OH	0,050	0,780	0,078
III	0,384	0,130	0,250				

† The roman numerals denote the oxygen atoms.

*Cuspidine-tilleyite [19], wollastonite [7], xonotlite [6], foshagite-hillebrandite-tobermorite [20], seidozerite [13], and lovenite [21].

**G.B. Bokii predicted silicon-oxygen patterns of this type in a course of lectures on chemical crystallography he presented at Moscow State University in 1955.

117

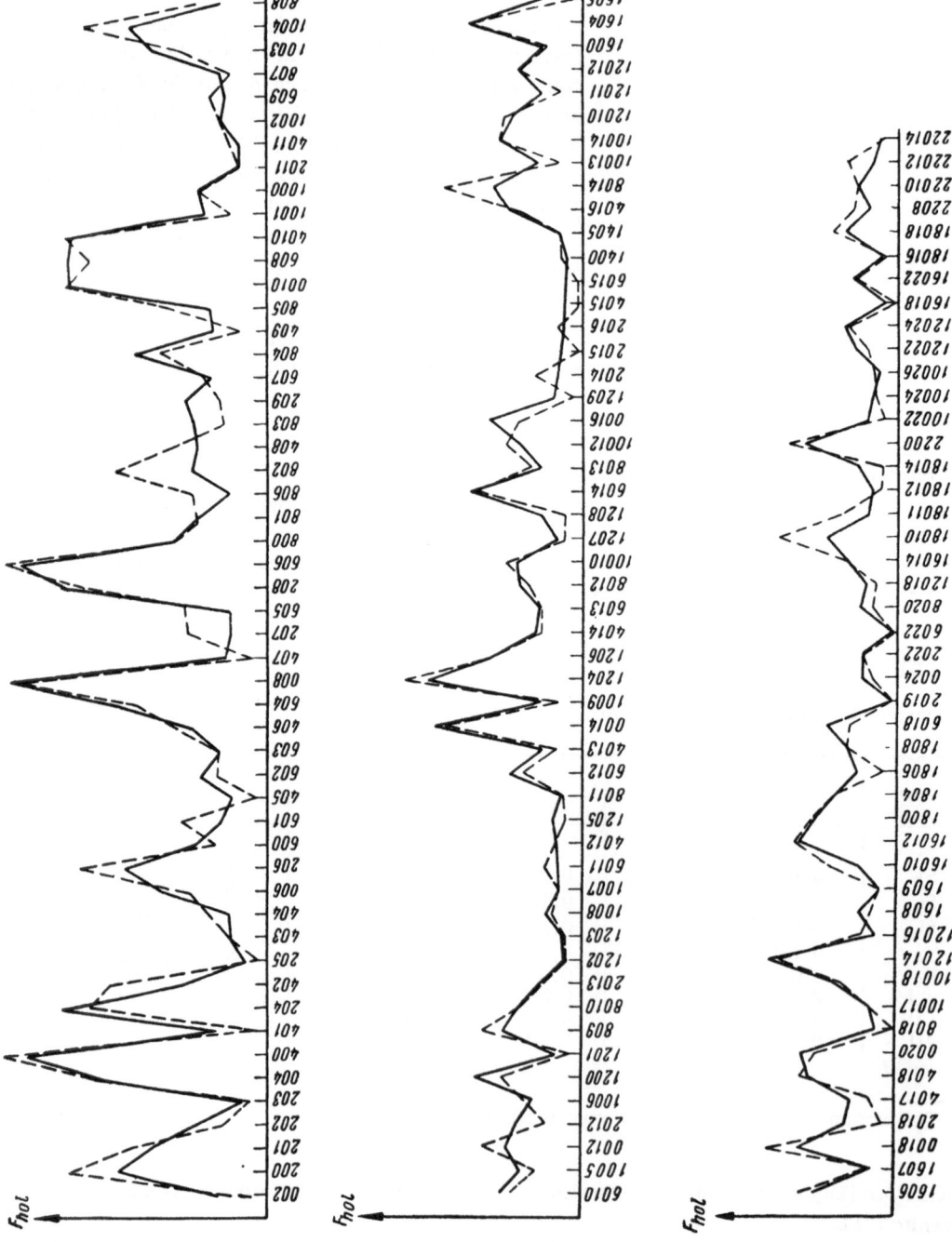

Fig. 6. Calculated and experimental $|F_{h0l}|$ (broken and full lines. respectively).

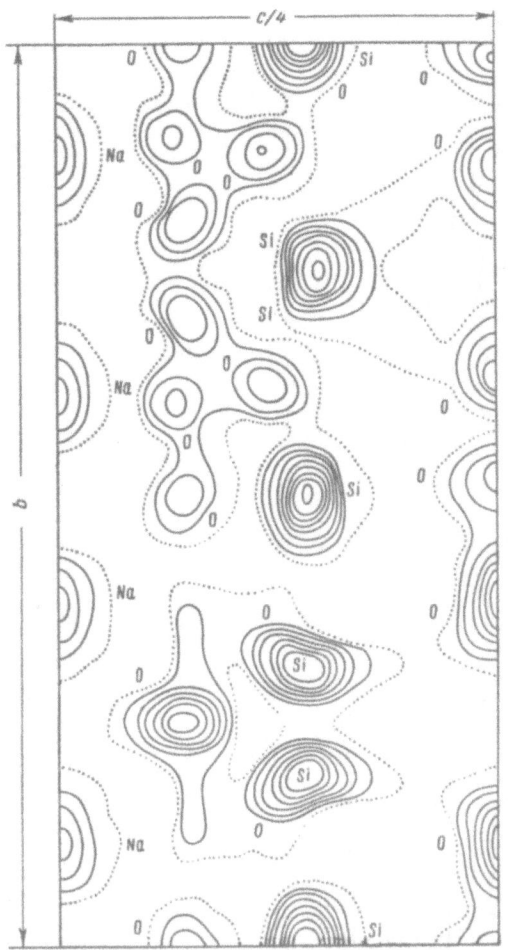

Fig. 7. Electron-density projection along $\underline{a}$.

strip, more exactly, $[Si_{4+2}O_{17}]$, epididymite has $2[Si_3O_9]-3O = [Si_6O_{15}] = [Si_{4+2}O_{15}]$. Figure 10 shows the wollastonite chain and the two kinds of strip produced by combining such chains. Figure 9b shows the $[Si_6O_{15}]_\infty$ strip, while Fig. 9c shows how this $[Si_6O_{15}]_\infty$ strip is combined with the twisted Na cubes. A Si_2O_7 group appearing in one half of the chain is linked to the diagonal square face of a Na polyhedron, while the third (single) Si tetrahedron is (see [17]) linked with the side of a square*. The second chain (half of the strip) is linked with the twisted cubes in another column.

A link of an epididymite chain divided by three is $[Si_2O_5]$, and a double chain having this unit for its links occurs in sillimanite $Al_2 SiO_5 = Al^{VI}[AlSi^{IV}O_5]$ [14], although in this case half of the tetrahedra contain Al instead of Si and the centers of gravity of all tetrahedra lie in one plane, so the link is actually $[Si_2O_5]$, or rather $[SiAlO_5]$. Figure 10 illustrates the differences between the two types of chain and the three types of strip (double chain). The combination of chains is controlled (as in the Ca silicates) by the need to match two large polyhedra and three tetrahedra in a distance of 7.3 A when we pass from the columns of Al octahedra in sillimanite [18] to those of Na polyhedra in epididymite, even though Si_2O_5 is still the basic formula. We can see that the Si-O-Si angle in the simple sillimanite Si_2O_5 strip is very much acute; the stress that would otherwise result is avoided by departure from crystallographic identity of all the Si(Al) atoms in a chain (that is, if we displace one third of them from the common plane, as in epididymite - Fig. 9b).

Epididymite shows a clear resemblance to the feldspars. Figure 11 shows the silicon-oxygen structure typical of orthoclase, which is very similar to the structure in epididymite, except that each link has four levels instead of three; $[Si_8O_{20}]_\infty = [Si_{4+4}O_{20}]_\infty$. Such strips may only be dissected from the three-dimensional framework formed by the tetrahedra in the case of the feldspars, whereas the $[Si_6O_{15}]$ links in epididymite are distinct and less tightly couple the single Be tetrahedra into layers; they leave room to fit the large Na atoms in the gaps between the strips (Figs. 12 and 13).

Epididymite resembles beryl and the aluminosilicates in having a framework similar to that found in SiO_2, because its formula corresponds to an (O+ OH):(Si+ Be) ratio of two. On the other hand, its structure is intermediate between the strip and framework structures found in silicates, because that structure contains dual $[Be_2Si_6O_{14}(OH)_2]$ layers consisting of $[Si_6O_{15}]_\infty$ strips and Be tetrahedra; epididymite must be called a beryllosilicate of layered structure in Bokii's classification of the silicates [23].

The doubled formula of epididymite $Na_2Be_2Si_6O_{14}(OH)_2$ bears to the formula $[Si_6O_{18}]$ a relation such that it is clear that one of the O atoms (out of 15) must be replaced by OH, the true formula being $[Si_6O_{14}OH]$. We cannot say which of the 15 atoms is so replaced, but it can hardly be one of the three (single) O atoms lying in the plane of symmetry, because these three are O atoms joined via two Si tetrahedra. The other 12 such atoms in that plane fall into six crystallographically identical pairs. The atoms O_{VII}, O_{VIII}, O_{IX} belong each to two Si tetrahedra at once, so one OH group is distributed statistically over the six positions of atoms O_I, O_{IV}, O_{VI} (compare the statistical distribution of two OH in afwillite [24] and of six OH

---

* This shortens one edge of the trigonal prism in which the Si_2O_7 group is inscribed and lengthens another (in favor of the single SiO_4 tetrahedron). Then the central O_{VIII} atom in the Si-O-Si group is displaced very much from the Si-Si axis (the Si-O-Si angle is 160°; see [22]).

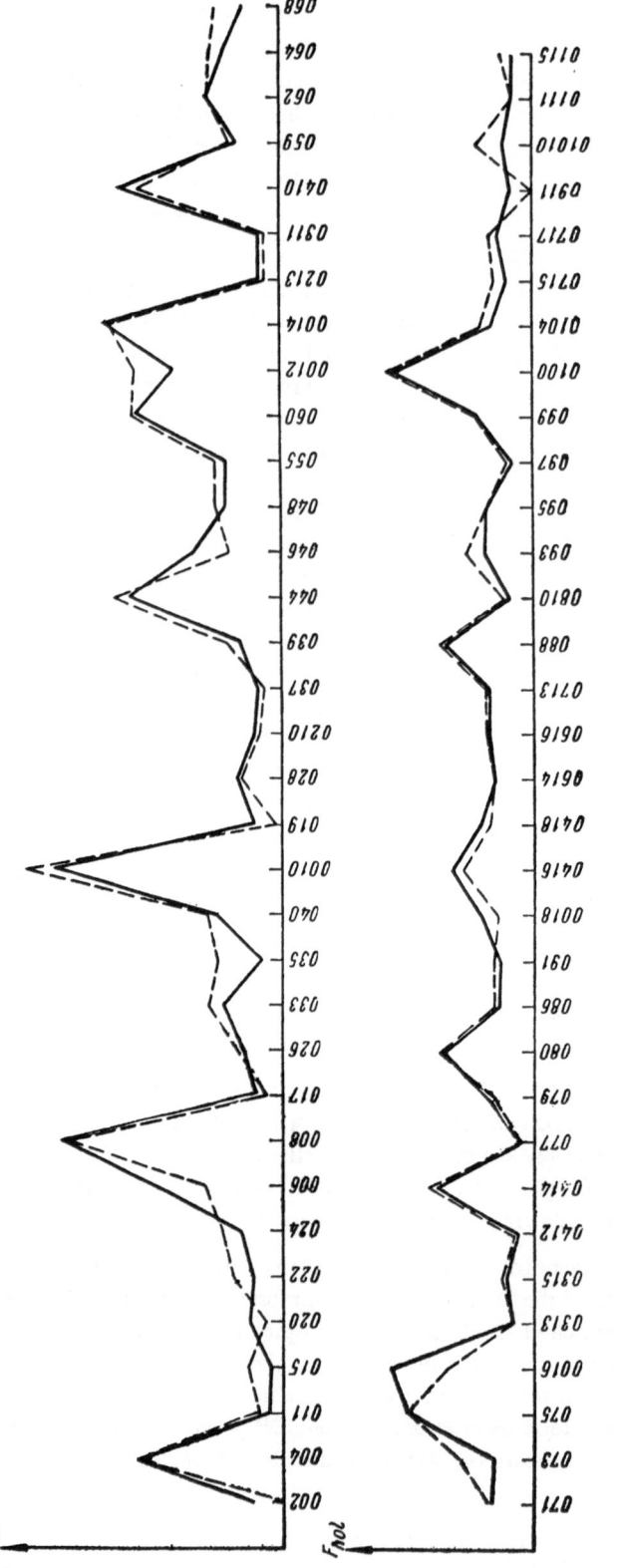

Fig. 8. The calculated and experimental $|F_{0kl}|$ (broken and full lines respectively).

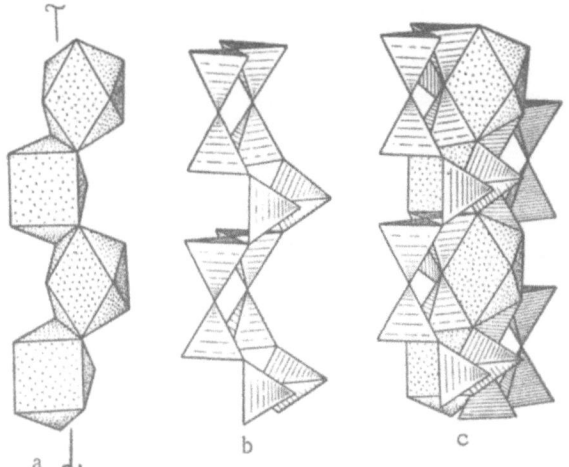

Fig. 9. Two basic patterns in epididymite: a) columns of Na polyhedra; b) double chain of Si tetrahedra; c) combination of the two patterns.

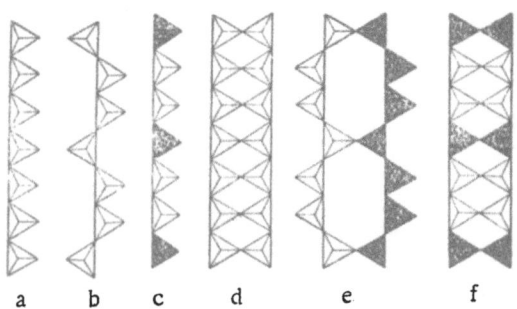

a   b   c   d   e   f

Fig. 10. Silicon-oxygen chains $[SiO_3]_\infty$ having their tetrahedra variously oriented, and some combinations thereof: a) simple $[Si_1O_3]_\infty$ chain; b) wollastonite $[Si_{2+1}O_9]_\infty$ chain (view along a normal to the plane containing the centers of gravity of the tetrahedra); c) the same seen from the side; d) sillimanite strip; e) xonotlite strip; f) epididymite strip.

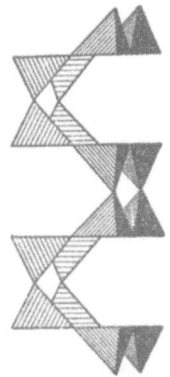

Fig. 11. $[Si_8O_{20}]_\infty$ strips in orthoclase.

TABLE 2

Interatomic Distances (Å) in Epididymite

Si_I- Tetrahedron		Be- Tetrahedron	
Si_I — II 1,60		Be — I 1,65	
Si_I — VI 1,85		Be — IV 1,68	
Si_I — VII 1,60		Be — VI 1,67	
Si_I — VIII 1,60		Be — OH 1,53	
II — VIII 2,78		I — IV 2,89	
II — VI 2,50		I — VI 2,42	
II — VII 2,52		I — OH 2,51	
VI — VII 2,66		IV — VI 2,87	
VI — VIII 2,56		IV — OH 2,56	
VII — VIII 2,43		VI — OH 2,51	
Si_{II}-Tetrahedron		Na- Polyhedron	
Si_{II} — I 1,59		Na — IV 2,11	
Si_{II} — III 1,66		Na — VI 2,50	
Si_{II} — VII 1,59		Na — VII 2,72	
Si_{II} — IX 1,69		Na — IX 2,20	
I — III 2,80		Na — VII 2,72	
I — VII 2,43		Na — IX 3,10	
I — IX 2,52		Na — VIII 2,08	
III — VII 2,76		Na — OH 2,63	
III — IX 2,42			
VII — IX 2,88			

Si_{III} Tetrahedron

Si_{III} — IV 1,63
Si_{III} — V 1,67
Si_{III} — VIII 1,62
Si_{III} — IX 1,74
IV — V 2,59

IV — VIII 2,90
IV — IX 2,53
V — VIII 2,84
V — IX 2,72
VIII — IX 2,50

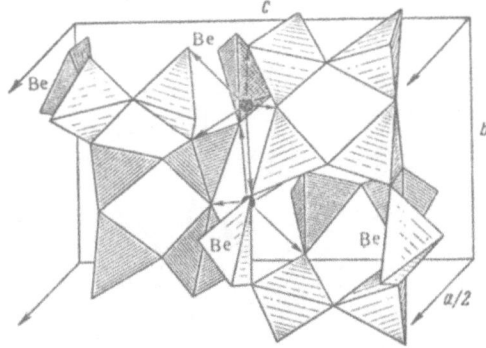

Fig. 12. Axonometric representation of the structure of epididymite.

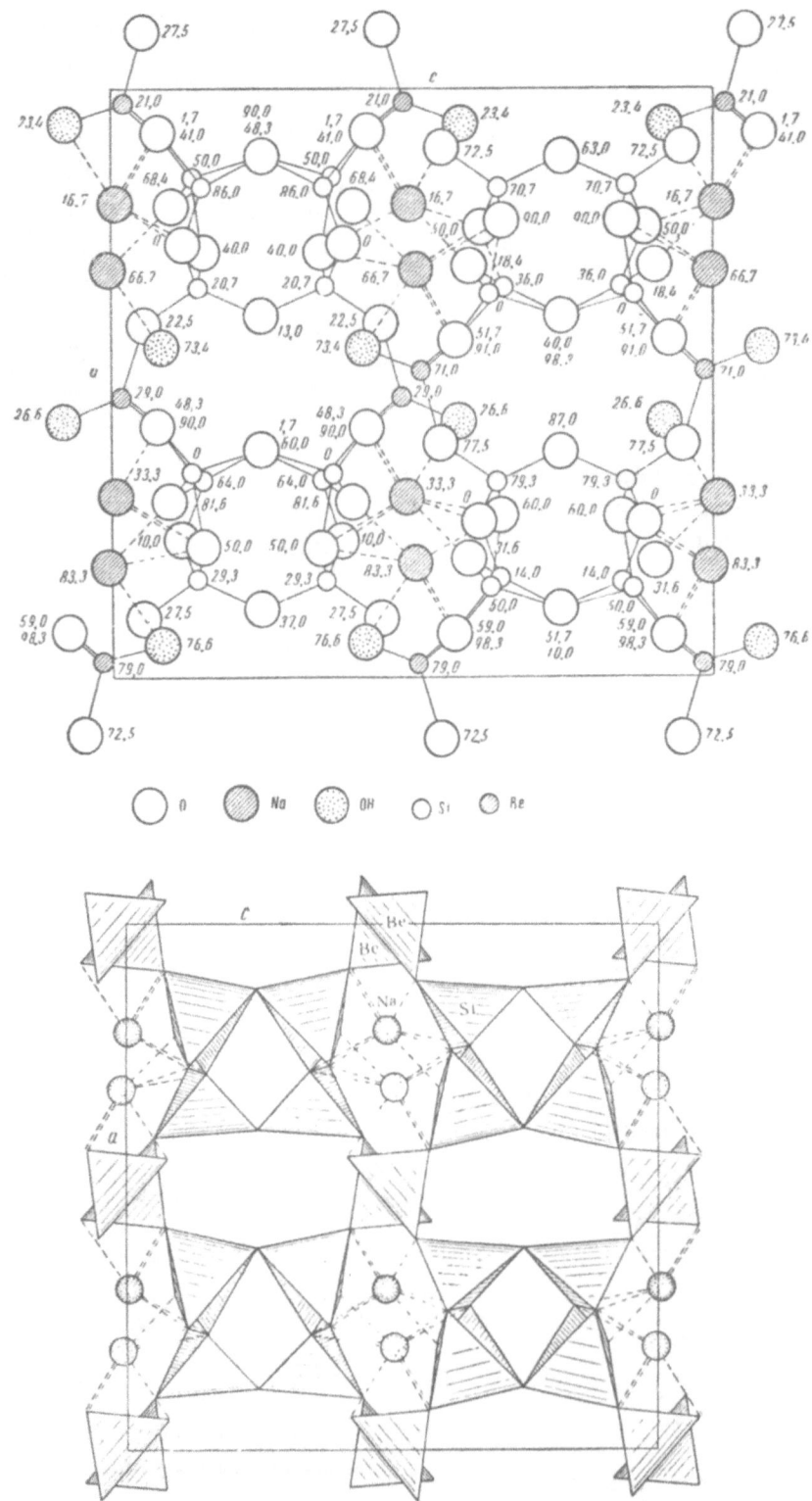

Fig. 13. Structure of epididymite : a) projection along <u>b</u> (the numbers are the heights in hundredths of <u>b</u>); b) xz projection in polyhedra.

TABLE 3

Interatomic Distances (A) in Ito's [4] Structure for Epididymite.

Si-Tetrahedron
Si — I 1,89
Si — III 1,13
Si — IV 2,38
Si — VI 1,79
I — III 2,84
I — IV 3,03
I — VI 2,64
III — IV 2,91
III — VI 2,80
IV — VI 3,01

Si- Tetrahedron
Si — II 1,89
Si — III 1,30
Si — V 2,38
Si — VII 1,79
II — III 2,84
II — V 3,03
II — VII 2,64
III — V 2,91
III — VII 2,98
V — VII 3,01

Si-Tetrahedron
Si — IV 2,86
Si — V 2,56
Si — VI 2,61
Si — VII 1,89
V — VII 3,01
V — IV 2,36
V — VI 2,91
VI — IV 3,01
IV — VII 2,91

Na- Octahedron
(centrosymmetric)
Na — III 2,91
Na — IV 2,20
Na — I 2,02
I — III 2,84
I — IV 3,03
III — IV 2,91

Na- Octahedron
(centrosymmetric)
Na — III 2,27
Na — II 2,03
Na — V 2,17
II — III 2,84
II — V 3,03
V — III 2,91

TABLE 4

Balance of Valencies in Epididymite

O (OH)	Na	Be	Si_I	Si_{II}	Si_{III}	Σ
I		$2/4$		$4/4$		$1\frac{1}{2}$
II			$2\times4/4$			2
III				$2\times4/4$		2
IV	$1/8$	$2/4$			$4/4$	$1\frac{5}{8}$
V					$2\times4/4$	2
VI	$1/8$	$2/4$	$4/4$			$1\frac{5}{8}$
VII	$2/8$		$4/4$	$4/4$		$2\frac{1}{4}$
VIII	$1/8$		$4/4$		$1/4$	$2\frac{1}{8}$
IX	$2/8$			$4/4$	$4/4$	$2\frac{1}{4}$
OH	$1/8$	$2/4$				$5/8$

in lovozerite [25]) * ; only the corners not coupled to Si atoms in the Be tetrahedra can for certain be assigned to hydroxyl.

The structure explains the physical properties of the mineral, especially the two mutually perpendicular cleavage planes (100) and (001); this feature makes epididymite very similar to orthoclase, the more so since the cause of the cleavage (silicon-oxygen strips extending along b) is the same. Along (001) we have closed double layers of Si tetrahedra with Be tetrahedra, the layers being separated by large Na atoms falling in a strictly hexagonal array; this feature makes the mineral similar to the micas, since tabular (pseudohexagonal) triplets tend to form, and so on** . The cleavage on (100) is not so good, because the silicon-oxygen strips are joined in that plane by Be tetrahedra.

Epididymite is optically positive, in accordance with the chain nature of the radical.

Eudidymite (the monoclinic form of epididymite) contains Si_6O_{15} chains arranged stepwise, but three tetrahedra still form one unit along the axis.

This form of chain explains the perfect cleavage on (110).

LITERATURE CITED

1. L.L. Shilin and E.I. Semenov, Doklady Akad. Nauk SSSR 122, 2, 325 (1957).***
2. B. Gossner and O. Kraus, Zbl. Mineral. A8, 257 (1927).
3. E.S. Dana, Textbook of Mineralogy [Russian translation] (Moscow, 1937).
4. T. Ito, Z. Kristallogr. 88, 142 (1934); X-ray Studies and Polymorphism (Tokyo, 1950).
5. H. Berman, Amer. Mineralogist 22, 391 (1937).
6. Kh.S. Mamedov and N.V. Belov, Doklady Akad. Nauk SSSR 104, 4, 615 (1955); Zap. Vses. Min. Obshch. 85, 1, 13 (1956).
7. Kh.S. Mamedov and N.V. Belov, Doklady Akad. Nauk SSSR 107, 3, 463 (1956).***
8. R.W.G. Wyckoff, The Structure of Crystals. Suppl. (New York, 1935).

*We cannot put the formula as $Na_2Be_2[Si_6O_{15}] \cdot H_2O$, because water is released only at temperatures above 800°C [1].

**Epitaxis of NaI (or other Na salts) on epididymite is reminiscent of the long-known epitaxis of KI on micas.

***Original Russian pagination. See C.B. translation.

9. G. Flink, Z. Kristallogr. 23, 353 (1894).

10. W.H. Zachariasen, Norsk. geol. tidsskr. 10, 449 (1929).

11. E.R. Howells, D.C. Phillips, and D. Rogers, Acta crystallogr. 3, 216 (1950).

12. Kh. S.Mamedov and N. V. Belov, Doklady Akad. Nauk SSSR 106, 3, 462 (1956).*

13. V.I. Simonov and N.V. Belov, Kristallografiya 4, 163 (1959).

14. M.A. Porai-Koshits, A Practical Course in X-ray Analysis of Structures [in Russian] (Izd. Mosk. Gos. Univ., 1960) Vol. II.

15. M.J. Buerger, Proc. Nat. Acad. Sci. USA 36, 272 (1951); Vector Space (New York, 1959) p. 239.

16. B.K. Vainshtein, Zhur. Éksp. i Teoret. Fiz. 27, 44 (1954).

17. N.V. Belov, "An outline of structural mineralogy. X," Min-Sborn. L'vov. Geol. Obshch. 13, 23 (1959).

18. N.V. Belov, Structures of Ionic Crystals [in Russian] (Moscow, 1947).

19. R.F. Smirnova, et al., Zap. Vses. Min. Obshch. 84, 1, 159 (1958).

20. Kh.S. Mamedov and N.V. Belov, Doklady Akad. Nauk SSSR 121, 5, 901 (1958);* 123, 1, 163 (1958)*; 123, 4, 741 (1958).*

21. V.I. Simonov and N.V. Belov, Doklady Akad. Nauk SSSR 130, 6, 1333 (1960).*

22. N.V. Belov, "A second chapter in the chemical crystallography of silicates," Zhur. Strukt. Khim. 1, 1, 39 (1960).*

23. G.B. Bokii, An Introduction to Chemical Crystallography [in Russian](Izd. Mosk. Gos. Univ., 1960) 2nd edition.

24. H.D. Megaw, Acta crystallogr. 5, 477 (1952).

25. V.V. Ilyukhin and N.V. Belov, Doklady Akad. Nauk SSSR 131, 1, 176 (1960).*

---

* Original Russian pagination. See C.B. translation.

Reprinted from Soviet Physics – Crystallography, Vol. 5, pp. 523-525, January-February, 1961

BAOTITE – A MINERAL WITH [Si$_4$O$_{12}$] METASILICATE RINGS

V. I. Simonov

Institute of Crystallography of the Academy of Sciences, USSR
Translated from Kristallografiya, Vol. 5, No. 4, pp. 544-546, July-August, 1960
Original article submitted March 31, 1960

The crystal structure of baotite (space group $I4_1/a$, dimensions of the unit cell: a = 19.68, c = 5.88 A) has been established by the method of minimalization of Patterson projections. The final atomic coordinates resulted from several computings of signs of hkl alternating with Fourier syntheses. Baotite is a new example of a ring silicate with a tetragonal radical [Si$_4$O$_{12}$]. The detailed formula of baotite is Ba$_4$(Ti, Nb)$_8$ClO$_{16}$[Si$_4$O$_{12}$]. The ring radicals [Si$_4$O$_{12}$] are on the 4-fold inversion axes, and (Ti, Nb)-octahedra are around 4-fold screw axes in columns analogous to those in the structure of rutile.

Recently in China a new silicate mineral, baotite, was discovered by the Soviet mineralogist Semenov and the Chinese scientist Hong Weng-hsing [1]. Two chemical analyses carried out by T. A. Kapitonova and A. V. Bykova led to the following formula for baotite: Ba(Ti, Nb)$_2$SiO$_7$ plus some Cl, approximately equal to one atom to four atoms of Si. The tetragonal cell of baotite (bipyramidal symmetry class) has the parameters: a = 19.68, c = 5.88 A. Its space group $C_{4h}^6 = I\,4_1/a$ is determined uniquely from the systematic extinctions of reflections. The central cell contains 16 formula units of the composition described above.

The determination of the crystal structure of baotite is facilitated by its high symmetry and by the presence of the heavy Ba atom (Z = 56). The horizontal glide planes of symmetry shorten by twofold the parameters a and b on the projection σ (x, y). The planar symmetry group of this projection is p4. Its independent unit is $^1/_{16}$ of the content of the body-centered baotite cell. On the projection of interatomic vectors p(x, y) it was possible to localize Ba. The Patterson peak corresponding to the vectors between the Ba atoms is linked with the second degree operation of the fourfold axis, and it serves as a basis for the construction of the (minimum) Buerger M-function [2]. The elementary operation of rotation about the fourfold axis allows one to obtain $M_4(x, y)$ from $M_2(x, y)$, which gives the base plane of the model of baotite. A more precise structure is obtained by alternate calculation of the sign of the structure amplitudes and projections of the electron densities. The final value of the reliability factor, calculated from all 164 values (Mo-radiation) of the structure amplitudes hk0 with $F_E \neq 0$ is equal to 15.3% with a maximum value of $\vartheta/\lambda = 1.02$. In F_C an isotropic temperature factor was introduced with Ba = 0.65 A.

In Fig. 1 is given a sketch of the baotite structure, all the cation polyhedra, except Ba, being shown. It is noticed that the fourfold metasilicate rings [Si$_4$O$_{12}$] are strung along the $\bar{4}$ axis. A similar ring was proposed by Ito [3] for the borosilicate axinite, but without strict proof. Strunz discussed the possibility of similar rings in neptunite [4], but the structure of this mineral still awaits a solution.

The geometrically analogous metaphosphate ring [P$_4$O$_{12}$] characterizes Al-metaphosphate Al$_4$[P$_4$O$_{12}$]$_3$ [5].

Between each pair of translationally identical rings [Si$_4$O$_{12}$] of baotite large cavities are formed in which the large Cl atoms are distributed at the general points of the fourfold inversion axis. Thus the solution of the structure elucidates the role and quantity of Cl (the absence of defects being assumed) in the mineral. The oxygen atoms of the silicate radical which form the vertical edges of the tetrahedra are simultaneously the

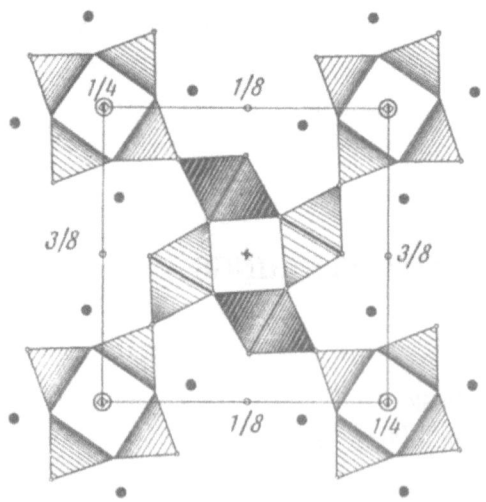

Fig. 1. Diagram of the structure of baotite. $^{1}/_{4}$ of the cell is shown. Ba) dark circles; Cl) light circles on the $\bar{4}$ axes.

The Horizontal Coordinates of the Basic Atoms of Baotite(along the a and b axes). Origin of Coordinates on the Fourfold Inversion Axis

Type of atom	x	y
Ba	0.153	0.029
$(Ti, Nb)_{I, II}$	0.137	0.220
Si	0.064	0.093
Cl	0.000	0.000
O_I	0.092	0.015
$O_{II, III}$	0.090	0.128
$O_{IV, V}$	0.195	0.184
$O_{VI, VII}$	0.081	0.244

apices of two (Ti, Nb)-octahedra. The central part of Fig. 1 (in which is shown one fourth of the whole tetragonal cell) is occupied by a pattern looking as if it had been excised from the rutile structure [1]. Four infinite vertical columns of (Ti, Nb)-octahedra, which are connected together by common horizontal edges, are close around the fourth-order screw axis (4_1) and form a continuous square tube. It will be noticed that the period (2.96 A) along the axis of the tube is equal to one octahedron,whereas in baotite (5.88 A) it is equal to two. This is a result of the different types of fourfold axes lying along the tube: 4_2 in rutile, 4_1 or 4_3 in baotite.

Expansion of the chemical formula of baotite to correspond to the crystal structure gives the formula $Ba_4(Ti, Nb)_8ClO_{16}[Si_4O_{12}]$. Four of these units (180 atoms) occur in the unit cell of the mineral. The Cl takes the parameter-less point in the special position of the $\bar{4}$ axis. The metasilicate rings $[Si_4O_{12}]$ lie in the plane perpendicular to this axis and level with its second special position. The peripheral O atoms of the silicate rings enter the (Ti, Nb)-octahedra and determine their vertical coordinates. The Ba atoms are placed in the large gaps in the structure at about the same height as the Cl.

The values of the horizontal coordinates of the basic atoms of baotite, obtained from the $\sigma(x, y)$ projection, are cited in the table. In the $\sigma(x, y)$ projection the atoms Ba, Si, Cl and O_I are not overlapped by other atoms. The remaining atoms of the structure are distributed in pairs along the z axis quite exactly one above the other, and in the table their horizontal coordinates are given the same value, though this is not required by symmetry. Exact fixing of all the individual coordinates of the basic atoms is hampered by the large value of the period a. The weighted projections along the c-axis constructed from reflections of any one of the layer lines are useless, because the distance between overlapping atoms is close to a half-period along Z.

The infinite colums of (Ti, Nb)-octahedra play a determining role in baotite. The length of the edges of these octahedra cause the structure to obey the rules of the "first chapter of silicate crystal chemistry" [7]. The one story $[Si_4O_{12}]$ rings are a consequence of the "commensurable" nature of the edges of the Si-tetrahedra and the edges of the (Ti, Nb)-octahedra.

We take the opportunity to thank N. V. Belov for discussion of the result of the present work and E. I. Semenov for the gift of monocrystals of baotite.

LITERATURE CITED

[1] E. I. Semenov and Hong Weng-hsing, Dizhi, keue, 10 (1959).

[2] M. J. Buerger, Acta Crystallogr. 4, 531 (1951).

[3] T. Ito and Y. Takeuchi, Acta Crystallogr. 5, 202 (1952).

[4] H. Strunz, Mineralogische Tabellen, 3 Auflage (Leipzig, 1957).

[5] L. Pauling and J. Sherman, Z. Kristallogr. 96, 481 (1937).

[6] N. V. Belov, The Structure of Ionic Crystals and Metallic Phases [in Russian] (1947) p. 77.

[7] N. V. Belov, Zhur. Strukt. Khim. 1, No. 1 (1960).

Reprinted from Soviet Physics—Crystallography, Vol. 5, pp. 540-548, January-February, 1961

CRYSTAL STRUCTURE OF NARSARSUKITE

Yu. A. Pyatenko and Z. V. Pudovkina

Institute of Mineralogy, Geochemistry, and Rarer-Element Chemical Crystallography
Translated from Kristallografiya Vol. 5, No. 4, pp. 563-573, July-August, 1960
Original article submitted March 10, 1960

The complete crystal structure investigation of narsarsukite $Na_2 (Ti, Fe) (O, OH)-[Si_4O_{10}]$ has been undertaken. It was found that narsarsukite is tetragonal, the unit cell dimensions being a = = 10.72, c = 7.99 A. Atomic arrangement was determined by Patterson and Fourier syntheses. Narsarsukite represents a new structure type with $[Si_4O_{10}]_\infty$ chains stretching along $\underline{c}$.

Narsarsukite is a rare titanosilicate found in alkali pegmatites in Greenland, the USA, and the USSR. Its composition is $Na_2(Ti, Fe)Si_4O_{11}$, with the Ti and Fe in the ratio 3:1 (by number of atoms, approximately); some (about $\frac{1}{2}$) of the O atoms are replaced by OH [1]. The mineral is tetragonal (Laue class 4/m) [1, 2] and the parameters are a = 10.80 and c = 8.01 A. The systematic absences correspond to the possible space groups $C_{4h}^5 - I4/m$, $C_4^5 - I4$ and $S_4^2 - I\bar{4}$. The morphology makes the first (centrosymmetric) group very probable [2], but difficulties have been encountered in further studies of the structure.

We have used specimens given to us by I. P. Tikhonenkov to record patterns from transparent fragments about 0.3×0.3 mm in cross section extended along $\underline{c}$ and terminated by perfect cleavage surfaces [on (100)].

The Laue class was confirmed as 4/m by means of an RKOP camera; the lattice parameters were found from oscillation photographs used in conjunction with pinacoid reflections recorded with an x-ray goniometer; they are: a = 10.72 ± 0.04, c = 7.99 ± 0.02 A. The density is 2.779 and the molecular weight is 384.3, so N = 3.99 ≈ 4.

A KFOR camera (retigraph) was used to record the O layer lines (by rotation on $\underline{c}$, and also on $0kl$). Rotation on $\underline{c}$ gave very good patterns, whereas that on $0kl$ gave rather few (about 60) spots, which were subject to very uneven absorption. The perfect cleavage on (001) prevented us from making isometric pieces for use in examination along $\underline{a}$.

The only spots were those corresponding to $h + k + l = 2n$. The result is that the x-ray group is 4/mI, $-/-$, i.e., is I4/m, I4 or I$\bar{4}$.

The atoms to be located in the cell are eight Na, 4(Ti + Fe), 16 Si, and 44 O. It is difficult to make any reasonable assumption about OH, but we may suppose that the Ti + Fe atoms are randomly distributed in one position. The Ti: Fe ratio is nearly 3:1, and these body-centered groups contain no position of multiplicity one or three.

The positions available in these space groups are such that the four Ti* atoms can lie only on the 4, $\bar{4}$, or 4_2 axes. It was difficult at this early stage to locate the other atoms (particularly on account of the choice of space groups), so we had recourse to analysis of the intensities. A 30-point blackening scale ($I_{min}: I_{max} = 1:300$) was used. The Lorentz and polarization factors were allowed for in calculating the F^2. The small size of the fragments (less than 0.3 x 0.3 mm) made absorption corrections unimportant.

First we constructed a Patterson projection along $\underline{c}$, which contained a few sharp peaks whose positions were such as to indicate that the heavy (Ti) atoms must be duplicated in the xy projection on one of the two

* Here and in future we denote (Ti, Fe) by Ti alone, for brevity.

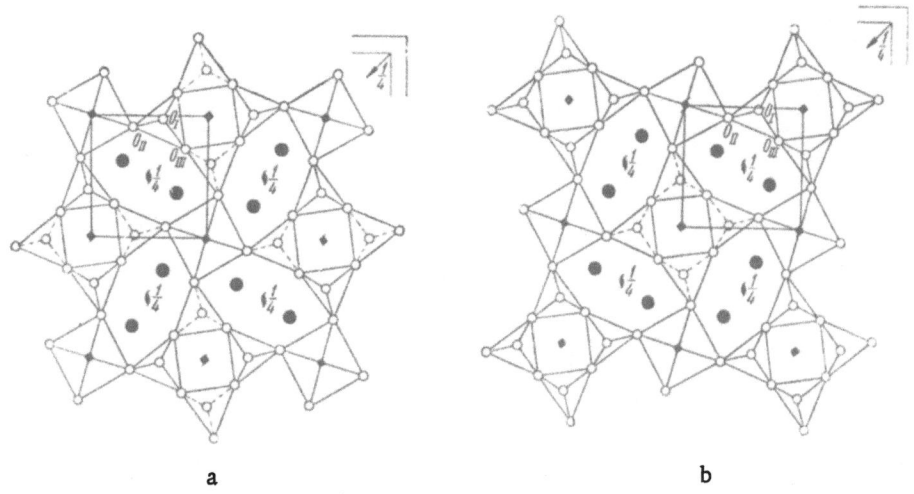

Fig. 1. Two possible models of the structure of narsarsukite.

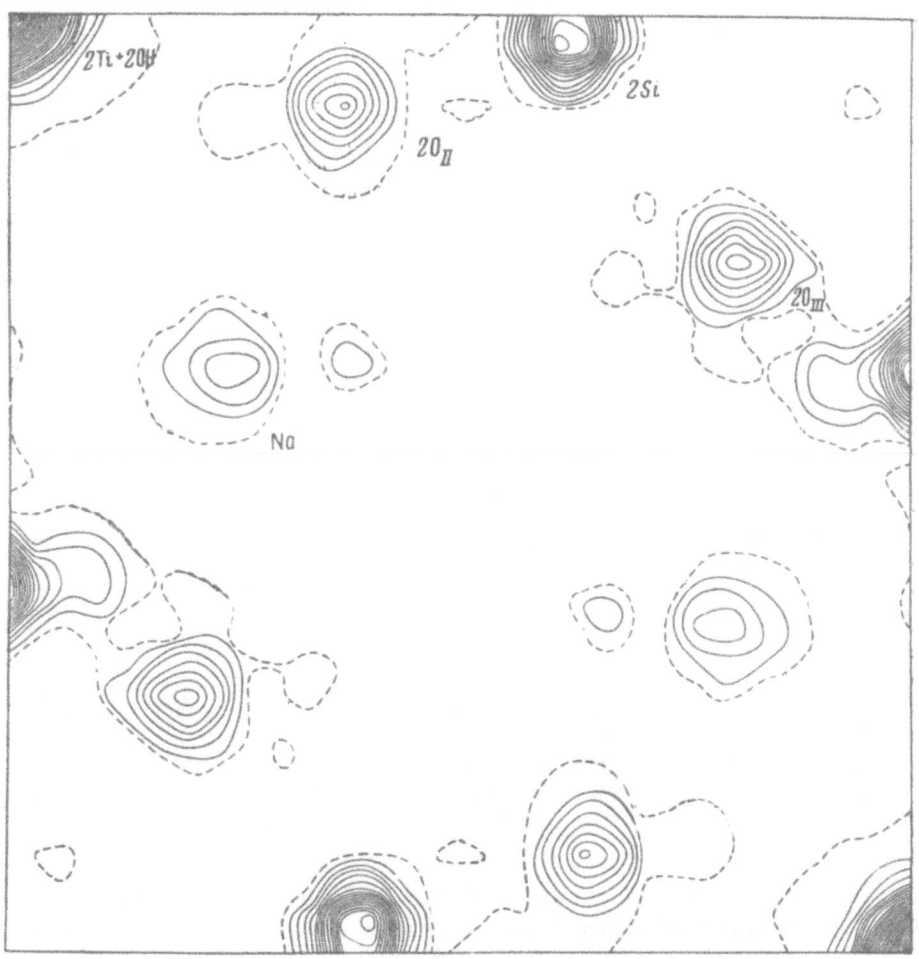

Fig. 2. Pattern (1/4 cell) for $\rho_2(xy)$.

sets of fourfold axes.   We took the (2Ti) atoms ($Z_{eff} > 44$) as lying at the origin of the xy projection, in order to get the peaks in P (uv) to represent the positions of most of the other atoms.   The only group giving a satisfactory interpretation of this is $C_{4h}^5 - I\,4/m$.   The main argument in favor of that group is that all the peaks

Fig. 3. Inclined projection $\sigma_2(xy)$. The negative regions are hatched.

TABLE 1

Comparison of the Horizontal Coordinates Derived from $\rho(xy)$ and $\sigma_2(xy)$

Atom	x		y		Coordinates adopted	
	$\rho(xy)$	$\sigma_2(xy)$	$\rho(xy)$	$\sigma_2(xy)$	x	y
Na	0.136	0.133	0.185	0.184	0.135	0.185
Si	0.307	0.307	0.013	0.013	0.307	0.013
O_I	—	0.025	—	0.299	0.025	0.299
O_{II}	0.180	0.175	0.049	0.049	0.178	0.049
O_{III}	0.401	0.402	0.133	0.134	0.402	0.134

in the xy projection can be explained only if we assume that all the atoms (not merely the Ti ones) are dupli-cated, i.e., that a horizontal reflection plane is present. Further evidence for group C_{4h}^5 was obtained sub-sequently.

The two possible positions of the Ti atoms (on the 4 or 4_2 axes in group I4/m) give rise to two models of the structure. Figure 1a shows the model with the Ti atoms on the rotation axes. The $\underline{x}$ and $\underline{y}$ coordinates used here derive directly from the Patterson projection, O_I excepted (O_I is not resolved from 2Si in this pro-jection, so the $\underline{x}$ and $\underline{y}$ for these atoms have been taken as the same). Figure 1a illustrates the three-dimen-sional network formed by Ti octahedra and rings of Si tetrahedra. The octahedra are joined at opposed vertices

T A B L E 2

Comparison of Measured and Calculated F_{hk0} (the F Include a Factor exp[− 0.56 × × (sin ϑ/λ^2]

hk0	F_0	F_0	hk0	F_0	F_0	hk0	F_0	F_0	hk0	F_0	F_0
020	125	+127	1.15.0	0	−5	5.13.0	49	+58	8.12.0	19	+8
040	83	+79	1.17.0	20	+32	5.15.0	13	−10	8.14.0	24	+24
060	191	+190	1.19.0	0	+6	5.17.0	27	+39	8.16.0	0	+15
080	23	+35	220	104	−109	620	143	+135	910	25	+21
0.10.0	121	+120	240	25	−23	640	94	+94	930	132	+124
0.120	24	+20	260	38	+25	660	94	+95	950	33	+37
0.14.0	17	+18	280	0	−6	970	65	+73	12.8.0	26	+30
0.16.0	76	+63	2.10.0	122	+129	990	47	+47	12.10.0	45	+45
0.18.0	0	+5	2.12.0	0	+5	9.11.0	49	+53	12.12.0	0	+1
110	36	+42	2.14.0	17	+23	9.13.0	73	+73	12.14.0	48	+56
130	128	+125	2.16.0	94	+79	9.15.0	24	+17	13.1.0	60	+63
150	10	−5	2.18.0	24	+31	9.17.0	31	+31	13.3.0	42	+52
170	15	+6	310	125	+125	10.2.0	36	+34	13.5.0	24	+35
190	20	−32	330	185	+218	10.4.0	107	+111	13.7.0	86	+71
1.11.0	0	−4	350	59	+66	10.6.0	72	+79	13.9.0	24	+30
1.13.0	39	+39	370	160	+170	10.8.0	64	+74	13.11.0	30	+37
390	79	+75	680	0	+18	10.10.0	44	+45	13.13.0	38	+40
3.11.0	53	+69	6.10.0	73	+81	10.12.0	0	−24	14.2.0	0	0
3.13.0	156	+121	6.12.0	55	+54	10.14.0	30	+44	14.4.0	49	+49
3.15.0	0	−5	6.14.0	51	+51	10.16.0	20	+16	14.6.0	18	−8
3.17.0	25	+19	6.16.0	42	+42	11.1.0	28	+32	14.8.0	0	−5
420	91	+81	6.18.0	26	+25	11.3.0	35	+49	14.10.0	36	+50
440	148	+145	710	138	+135	11.5.0	26	−47	14.12.0	0	+12
460	136	+118	730	124	+116	11.7.0	16	+24	15.1.0	0	+5
480	11	−9	750	12	−25	11.9.0	0	0	15.3.0	0	−7
4.10.0	35	+37	770	97	+98	11.11.0	0	−9	15.5.0	23	+32
4.12.0	0	+4	790	25	+27	11.13.0	20	+25	15.7.0	57	+46
4.14.0	0	+7	7.11.0	0	+7	11.15.0	0	−14	15.9.0	0	+4
4.16.0	0	+4	7.13.0	29	+36	12.2.0	20	+31	15.11.0	25	+21
4.18.0	20	−8	7.15.0	0	+2	12.4.0	42	+45	16.2.0	0	+16
510	95	+96	7.17.0	28	+46	12.6.0	14	−11	16.4.0	95	+79
530	0	−9	820	73	−92	16.6.0	30	+35	17.7.0	0	+9
550	98	−106	840	0	+10	16.8.0	29	+32	17.9.0	0	+9
570	140	+145	860	18	−2	16.10.0	58	+49	18.2.0	26	−22
590	0	−7	880	22	−23	17.1.0	77	+63	18.4.0	20	+26
5.11.0	27	−40	8.10.0	41	+62	17.3.0	27	−32	18.6.0	17	−4
						17.5.0	0	+12	19.1.0	29	+37

into chains along z; the c period equals the length of two links in the chain. The rings are formed by SiO_4 tetrahedra facing in opposite directions in pairs. The rings are coupled vertically via their O_I atoms into infinite $[Si_4O_{10}]_\infty$ chains or columns. The circles in Fig. 1 denote the Na atoms.

In the second model (Fig. 1b) the Ti atoms lie on the 4_2 axes. The structure follows the same general plan, the most important difference being that the tetrahedra take up other positions in the rings; all four are oriented identically. These different rings are coupled in a different fashion; in the first model the rings formed chains, whereas in the second we have duplicated four-member rings having the same Si : O ratio in their radicals.

The two models have identical P(uv), as follows from the fact that F^2_{hkl} (I) = F^2_{hkl} (II) for hk$l$ reflections having l even [3]. Later we analyzed the odd layer lines and proved conclusively that the first model represents the true structure, but even at this stage there were very strong indications that the first model was the correct one; the distances in the structures provided valuable evidence. In both models there are tetrahedra nearly regular in shape and so oriented that their heights are parallel to z. Here c = $4h_{tetr}$, so a is $\sqrt{6/8} \cdot c = 2.45$ A. A tetrahedron of side 2.45 A is improbable; some rotation (distortion of the four-membered rings) is required in order to produce normal distances. The axis of the rings is a 4 axis in the second model, and that axis allows only one type of distortion, namely rotation of the tetrahedra about the line joining adjacent O_{III} atoms; the O_{III}−O'_{III} distance remains low. But if the axis is 4_2 (first model), the symmetry allows a greater variety of rotations, which means that normal spacings can be achieved.

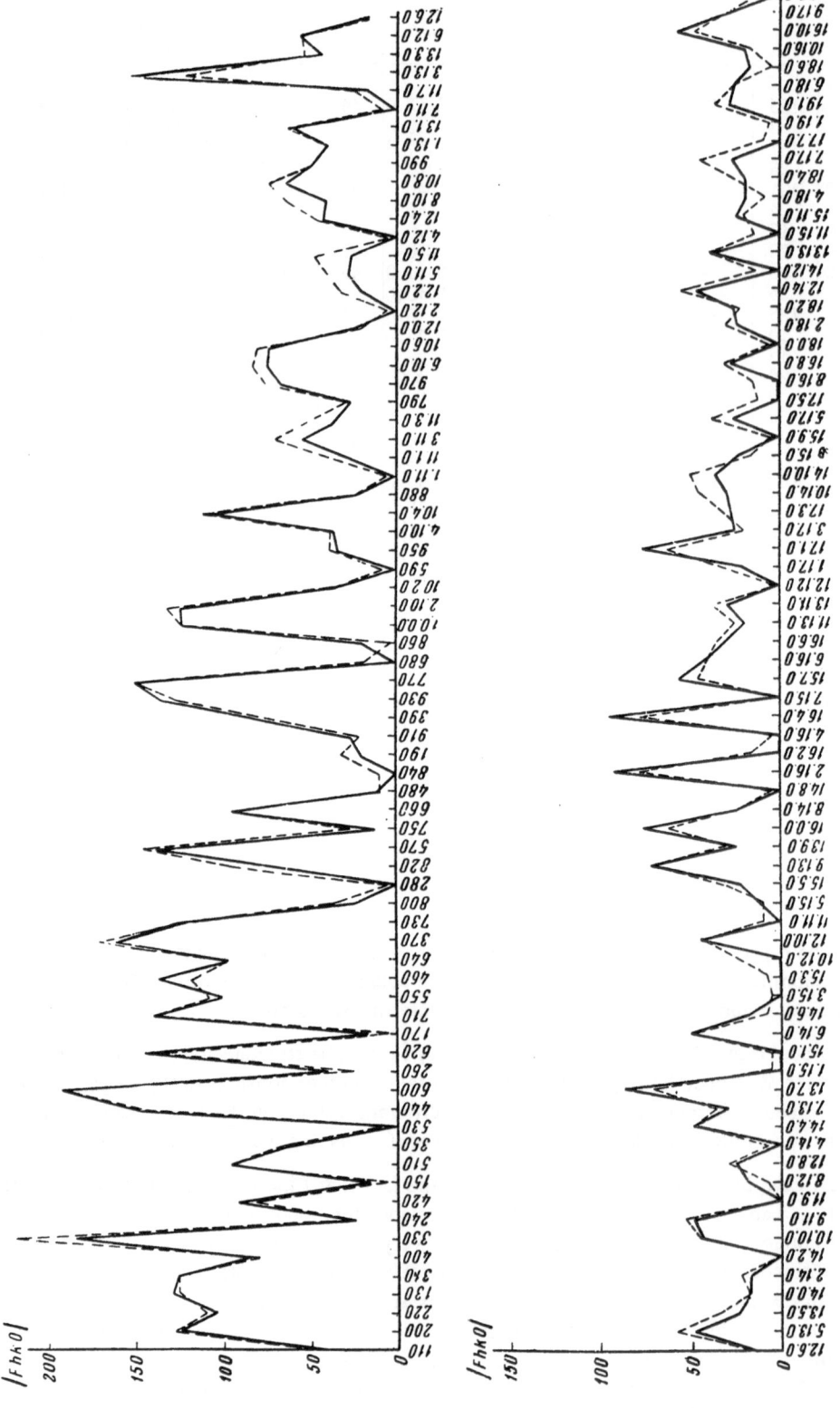

Fig. 4. Comparison of experimental (————) and calculated (– – – –) F_{hk0}.

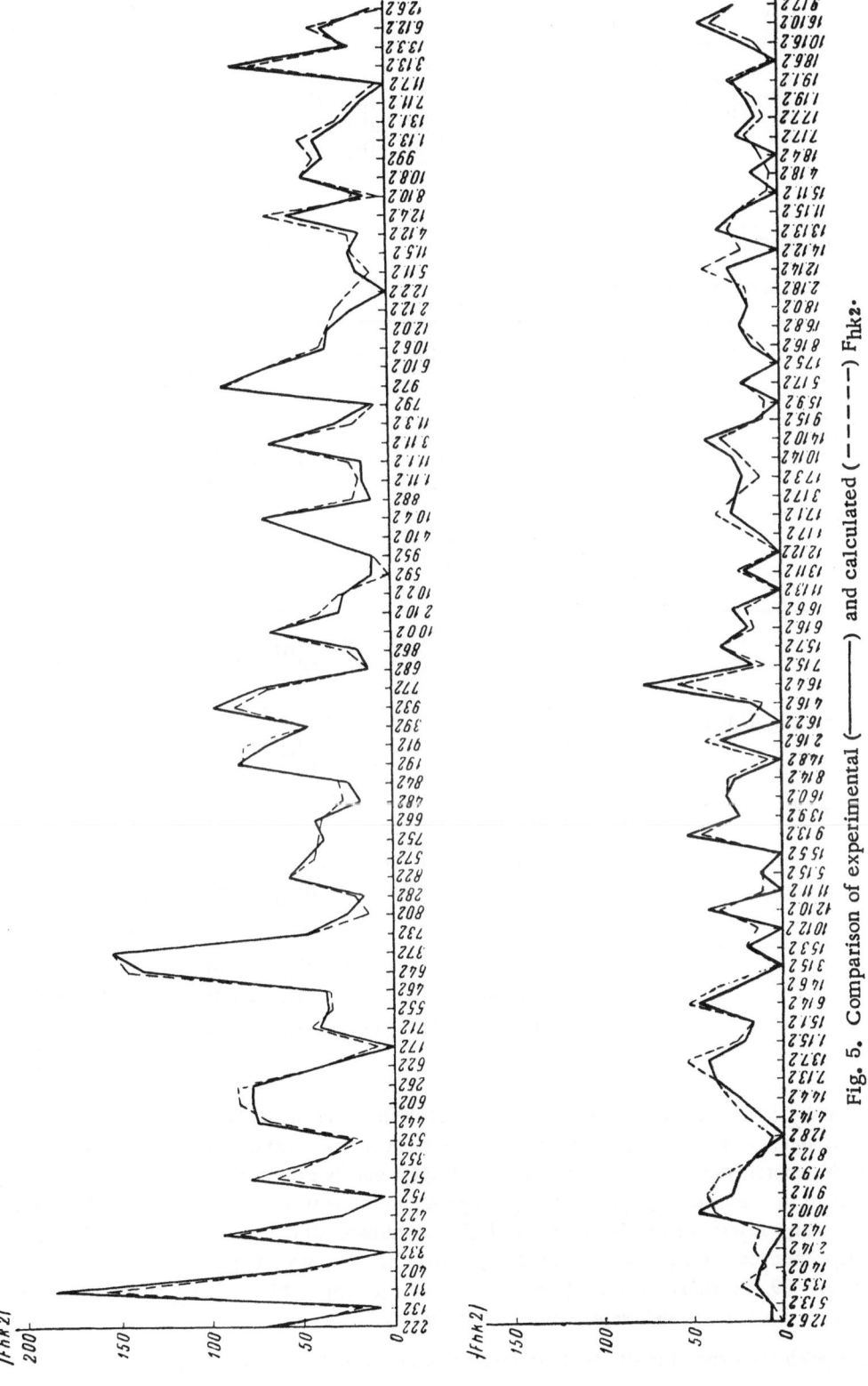

Fig. 5. Comparison of experimental (———) and calculated (– – – –) F_{hk2}.

TABLE 3

Atomic Coordinates for Narsarsukite

ATOM		x	y	z	ATOM		x	y	z
Na	(8)	0.135	0.185	(0)	O_{II}	(16)	0.178	0.049	0.225
(Ti, Fe)	(4)	(0)	(0)	0.250	O_{III}	(16)	0.402	0.134	0.302
Si	(16)	0.307	0.013	0.298	$(O, OH)_I$	(2)	(0)	(0)	(0.500)
O_I	(8)	0.025	0.299	(0.500)	$(O, OH)_{II}$	(2)	(0)	(0)	(0)

TABLE 4

Interatomic Distances in Narsarsukite

Polyhedron	Interatomic distances		Limits of variation
1	2	3	4
Ti-octahedron	Ti — $(O, OH)_I$ = 2.00 Ti — $(O, OH)_{II}$ = 2.00	Ti — O_{II} = 1.98	1.98—2.00
	O_{II} — O'_{II} = 2.80 O_{II} — $(O, OH)_{II}$ = 2.68	O_{II} — $(O, OH)_I$ = 2.96	2.68—2.96
Si-tetrahedron	Si — O_I = 1.67 Si — O_{II} = 1.55	Si — O_{III} = 1.65 Si — O'_{III} = 1.57	1.55—1.67
	O_I — O_{II} = 2.67 O_I — O_{III} = 2.58 O_I—O'_{III} = 2.64	O_{II} — O_{III} = 2.64 O_{II} — O'_{III} = 2.57 O_{III} — O'_{III} = 2.65	2.57—2.67
Na-octahedron	Na — O_{II} = 2.36 Na — O'_{II} = 2.67	Na — O_{III} = 2.54 Na — $(O, OH)_{II}$ = 2.46	2.36—2.67
	O_{II} — O'_{II} = 2.80 O_{II} — O''_{II} = 3.60 O_{II} — $(O, OH)_{II}$ = 2.68	O'_{II} — O''_{III} = 2.57 O_{III} — O''_{III} = 3.51 O_{III} — O'''_{III} = 3.16	2.57—3.60

The x and y of the first model were used to establish the signs of the F_{hk0} (identical coordinates were adopted for O_I and Si). We used 135 independent reflections in the Fourier synthesis for $\rho_1(xy)$; this synthesis showed that the coordinates of Na had been chosen wrongly, but that the model was otherwise reliable. Fresh coordinates were used to find the signs of the F_{hk0} again and to construct $\rho_2(xy)$ (Fig. 2). Although the pattern was very sharp, it was not possible to establish the coordinates of O_I; the only useful point was that the peak was unsymmetrical, so O_I does not lie exactly above Si. The peak heights (in el /A^2) are ρ (2Ti + 2OH) = = 229.6, ρ (Na) = 29.4, ρ (2Si) = 99.8, ρ ($2O_{II}$) = 47.2 and ρ ($2O_{III}$) = 51.6. The x and y were used to calculate theoretical F_{hk0} (B = 0.6 A^2), and were compared with the F_0.

Here R (with allowance for all visible reflections up to $\sin\theta/\lambda \leq 1.1$) was 17.1%.

Next we used P(vw) in an attempt to find the z coordinates. The patterns to be used contained only about 60 spots, so we could not expect to get a sharp pattern by projection along the long (10.72 A) a axis.

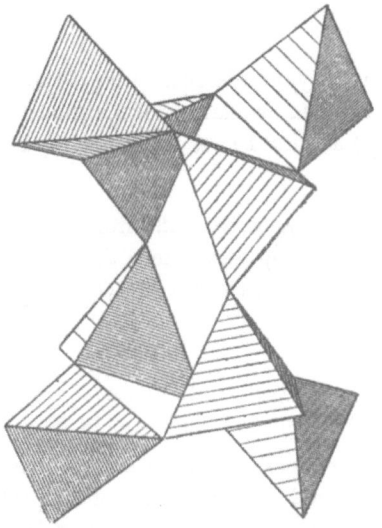

Fig. 6. Two links in the $[Si_4O_{10}]_\infty$ chains.

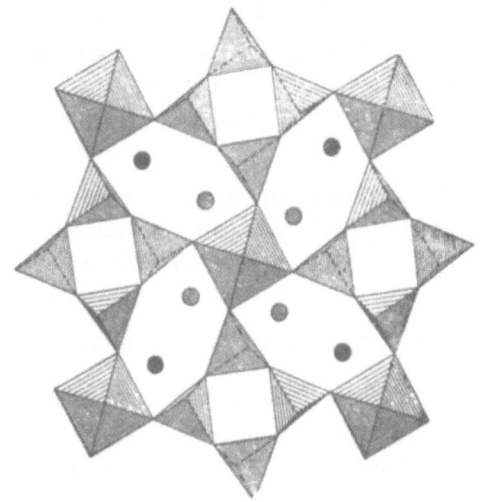

Fig. 7. Projection of the structure of narsarsukite along $\underline{c}$. (The circles denote Na atoms at various heights.)

TABLE 5

Balance of Valences in Narsarsukite

Atom	Si	Ti + Fe	Na	Sum
O_I	$2 \cdot {}^4/_4$	—	—	2
O_{II}	${}^4/_4$	${}^{15}/_4 : 6$	$2 \cdot {}^1/_7$	$2 - {}^5/_{56}$
O_{III}	$2 \cdot {}^4/_4$	—	${}^1/_7$	$2 + {}^1/_7$
$(O, OH)_I$	—	$2 \cdot {}^{15}/_4 : 6$	—	$1 + {}^1/_4$
$(O, OH)_{II}$	—	$2 \cdot {}^{15}/_4 : 6$	$4 \cdot {}^1/_7$	$2 - {}^5/_{28}$

Inclined Projections

A projection $\int P(uvw) \cos 2\pi\, 2w\, dw$ was constructed from 140 independent hk2 reflections to test the model and to bring the hk0 and hk2 into correspondence, which is difficult in class 4/m because $F^2_{hkl} \neq F^2_{\bar{h}kl}$. The results confirmed that the model is correct; this ideal model was then subjected to analysis in order to establish the vertical coordinates more exactly (the $\underline{x}$ and $\underline{y}$ being already known).

All the spacings were brought into agreement with the standard ones quite easily; the second model did not allow of this.

The signs of the F_{hk2} were determined from the $\underline{x}$ and $\underline{y}$ derived from the final $\rho\,(xy)$ and from the $\underline{z}$ derived from the model and corrected in accordance with the standard spacings. The projection

$$\sigma_2\,(xy) = \int\limits_0^1 \rho\,(xyz) \cos 2\pi\, 2z\, c\, dz$$

constructed from the F_{hk2} (Fig. 3) corresponds well with the model and with the previous syntheses. The most important result obtained from $\sigma_2\,(xy)$ was the $\underline{x}$ and $\underline{y}$ coordinates of O_I, which were not obtainable from $\rho\,(xy)$. Table 1 gives a comparison of the horizontal coordinates derived from $\rho\,(xy)$ and $\sigma_2\,(xy)$.

The new $\underline{x}$ and $\underline{y}$ were used to calculate theoretical F_{hk0}, which were compared with the F_0 (Table 2 and Fig. 4); the results are R = 13% (for $F_0 \neq 0$) and R = 17.4% for all reflections up to $\sin\theta/\lambda \leq 1.1$.

Coordinates and Atomic Separations

The $0kl$ line had few spots (the axis is 10.72 A long), but the ρ (yz) constructed from 60 spots gave all coordinates except the $\underline{z}$ of O_{II}; the values are very close to those found by analysis of the model. Figure 5 compares the two sets of F_{hk_2}. The results are R = 16.5% ($F_0 \neq 0$) and R = 19.3% for all hk2 up to $\sin\theta/\lambda \leq \leq 1.1$.

The coordinates finally adopted are given in Table 3; here $\underline{z}$ for O_{II} derives only from chemical crystallography. Four O atoms partly replaced by OH lie in positions not having parameters; the Ti atoms are defined by one parameter (00z), the Na and O_I by two parameters (xy0), and the rest (Si, O_{II}, O_{III}) are in general positions. In all, there are 14 parameters (Table 3).

Table 4 gives the interatomic distances.

Description of the Structure and Balance of Valences

The structure is based on chains of (Ti, Fe) octahedra joined by opposite vertices and on chains $[Si_4O_{10}]_\infty$ extending along $\underline{c}$. The members of the latter chains are four-membered rings composed of SiO_4 tetrahedra joined at their corners in pairs facing in opposite directions (Fig. 6). The (Ti, Fe) chains lie on fourfold rotation axes, while the axes of the $[Si_4O_{10}]_\infty$ chains are 4_2 axes. Horizontally the two kinds of chain are joined at the free vertices of octahedra and tetrahedra (Fig. 7); the octahedra and tetrahedra are almost regular. The coordination polyhedron of Na has seven corners, being a triangular prism having a pyramid on one of its faces. Polyhedra of this type have been found in ilvaite [4] and other minerals.

Table 5 gives the balance of valences, for which purpose the valence + 15/4 has been assigned to the cations at the centers of the octahedra (Ti: Fe $\approx$ 3:1).

The valences indicate that the OH ions are distributed over the two twofold positions of group I4/m, with a considerable preference for position 2(b): $00\frac{1}{2}$.

The properties of narsarsukite correspond well with the structure. The mineral is optically positive, and the birefringence is high (0.020-0.040); these features correspond to the chains of (Ti, Fe) octahedra and chains of silicon-oxygen figures. The large holes in the structure result in a low specific gravity (2.78). The cleavage planes pass through the O atoms at points where the chains of both sorts are joined.

The structure can be considered as of a new type. Other structures having the $[Si_4O_{10}]$ radical are layered (mica, apophyllite), but in narsarsukite that radical forms chains extending indefinitely in one dimension (compare $[Si_2O_5]$ in sillimanite).

The minerals most similar in structure to narsarsukite are those of the scapolite group, but in those minerals it is scarcely correct to separate four-membered rings, because the scapolites are silicates having extended frameworks rather than chains or rings.

LITERATURE CITED

[1] B. Gossner and H. Struns, Z. Kristallogr. 82, 150 (1932).

[2] B. E. Warren and C. R. Amberg, Amer. Mineralogist. 19, 546 (1934).

[3] International Tables for X-Ray Crystallography 1 (1952).

[4] N. V. Belov and V. I. Mokeeva, Trudy Inst. Kristall. Akad. Nauk SSSR, 9, 47 (1954).

Reprinted from Soviet Physics—Crystallography, Vol. 5, pp. 659-667, March-April, 1961

CRYSTAL STRUCTURE OF SPURRITE

R. F. Klevtsova and N. V. Belov

Institute of Crystallography, Academy of Sciences of the USSR;

Institute of Inorganic Chemistry, Siberian Division, Academy of Sciences of the USSR

Translated from Kristallografiya, Vol. 5, No. 5, pp. 689-697, September-October, 1960

Original article submitted May 21, 1960

The unit cell parameters of spurrite were given in 1955 in [1] as: $a = 10.46$, $b = 6.70$, $c = 14.20$ A, $\beta = 101° 19'$. The wide-spread monoclinic space group $C_{2h}^5 = P2_1/a$ characterizes the Ca silicates related to spurrite: tilleyite $Ca_5[Si_2O_7][CO_3]_2$ [2], cuspidine $Ca_4[Si_2O_7]F_2$ [3], scotite $6CaSiO_3 \cdot CaCO_3 \cdot 2H_2O$ [4]. There are four molecules of $Ca_5Si_2O_8CO_3$ in the unit cell ($\delta = 3.01$).

Twelve Weissenberg layer-line patterns (layers 0-5) were obtained by rotation of the crystal around the shorter a and b axes (Mo radiation); 1492 (axis a) and 2701 (axis b) reflections were recorded from them, of which approximately 3500 were independent.

The reflection intensities were evaluated visually from a blackening scale, and for each of them (first on a conventional relative scale) a unitary structure amplitude [5] was calculated.

The majority of the strong reflections have indices with $(h) + k = 2n$, i.e., the structure has a pseudo-lattice-constant $b/2$, which occurs more pronouncedly in the structure of cuspidine, and also in wollastonite [6] and in xonotlite [7].

From the F^2 $(h0l)$ values the Patterson projections $p(xz)$ and $p(yz)$ were constructed, but since there was a large number (76) of atoms in the cell occupying, apparently, only the general positions, these syntheses did not yield direct conclusions as to the structure, and we turned to direct methods of interpretation, in particular, to the method of juxtaposition, which was so well justified in interpreting the structure of cuspidine [3].

This deciphering of cuspidine was begun by the juxtaposition method, the working plan of which was set forth by Rumanova [8, 9]. It is most convenient in the case of monoclinic groups with a glide plane to use the xy and yz projections which are normal to this plane, and which allow one, by juxtaposition of corresponding reflections, to utilize the relationships easily derived from the expanded expression of the structure factor for the group $P2_1/a$ [10]:

$$
\begin{aligned}
\left. \begin{array}{l} F_{hkl} = F_{h\bar{k}l} = F_{\bar{h}\bar{k}\bar{l}} = F_{\overline{hkl}} \\ F_{\bar{h}kl} = F_{hk\bar{l}} \neq F_{hkl} \end{array} \right\} \quad & \text{for} \quad h + k = 2n, \\
\left. \begin{array}{l} F_{hkl} = -F_{h\bar{k}l} = -F_{\bar{h}k\bar{l}} = F_{\overline{hkl}} \\ F_{\bar{h}kl} = -F_{hk\bar{l}} \neq F_{hhl} \end{array} \right\} \quad & \text{for} \quad h + k = 2n + 1 .
\end{aligned}
$$

According to the working scheme [8, 9], of 200 $0kl$ reflections (i.e., from the layer line corresponding to the smaller axis a) 55 of the strongest reflections (approximately 27%) were separated out into a "supporting group", and similar supporting groups were created from the other layer lines on a. Because of the peculiarity — a pseudo lattice-constant along the b axis — of 55 reflections gathered into the basic group only four had indices with $(h) + k = 2n + 1$, and the majority of the comparisons carried out had to do with the group of reflections with $(h) + k = 2n$ (see [8, 9]).

For each $k_1 \neq k_2$ and $l_1 = l_2$ we had 38 comparisons; moreover, each was proved by 20-25 determining pairs (see [8, 9]), among which there were amplitudes from nonzero layers. The comparison of the $0kl$ amplitudes with $k_1 = k_2$ and $l_1 \neq l_2$ was carried out with very limited data (determined on one zero layer), which gave very weak ratios strengthened by 3-6 pairs only. In order to determine the signs of all the amplitudes, it was necessary to expand the supporting group by the introduction of reflections with $\underline{k}$'s odd, so that their number reached 19 (but judging from the values of the unitary structure amplitudes they were much smaller than those in 25% of the supporting ones).

As was to be expected, it was not possible to include in the determining pairs amplitudes with $(h) + k = 2n + 1$, and therefore (in accordance with the space group dependence for the group $P2_1/a$) our pairs, when compared, always showed only a similarity of signs, which decreased the reliability and the value of the determination. Nevertheless, as a result of comparison, the reflections of the supporting $0kl$ group broke up into six subgroups of 4-10 reflections each with signs conventionally indicated by the next letter of the alphabet.

The subsequent reduction in the number of letters was performed by a purely statistical method within the supporting group, and this was the only possible approach to the signs of the amplitudes when $k = 2n + 1$; only four pairs could be selected from among them, and two actually showed a contrast in signs:

$$
\begin{array}{llll}
0.\ 1.\ 6 & 0.\ 11.\ 6\ \} & \overline{24}\ \overset{+}{0} \\
0.\ 5.\ 7 & 0.\ 1.\ 7\ \} & \overset{+}{22}\ \overline{10} \\
0.\ 5.\ 11 & 0.\ 11.\ 11\} & \overset{+}{5}\ \overline{0} \\
0.\ 3.\ 14 & 0.\ 1.\ 14\} & \overline{27}\ \overset{+}{3}
\end{array}
$$

The final result of the analysis was to reduce the signs of the $0kl$ reflections with $\underline{k}$ even to three letters: $\underline{a},\ \underline{c},\ \underline{e}$. One of them,* according to the rules for primitive projections (see [9, 11] and especially [12]), can be taken as +; then, for reflections with $k = 4n$, we have either + or $\underline{c}$; and for those with $k = 4n + 2$** either $\underline{e}$ or ce.

The sign of $\underline{e}$ was determined when the signs of the $h2l$ reflections were established, which first were expressed with the aid of the $h0l$ and $h2l$ reflections (in which $\underline{e}$ was included), and then of $h1l$ reflections. The equalities thus obtained gave a + sign in 41 cases out of 47; for example,

$$
\left.
\begin{array}{l}
S\,(4.\ 2.\ \overline{10}) = \bar{e}\ \{\bar{7}\} \\
S\,(4.\ 2.\ 8\ ) = \bar{e}\ \{\bar{2}\} \\
S\,(6.\ 2.\ \overline{15}) = e\ \{\overset{+}{9}\}
\end{array}
\right\} e = +
$$

On the basis of the $0kl$ supporting group thus established, the signs of the remaining $0kl$ reflections were found by a purely statistical method. The sign of the 31st reflection was not determined (15.4%). From the amplitudes with signs a synthesis of the electron density $\sigma\,(yz)$ was constructed, on which, because of the circumstances just discussed,*** the majority of the peaks were diffuse, nevertheless, there were some very strong ones on the levels $y = 0$, $1/4$, $1/2$, $3/4$, which must correspond to the Ca and Si atoms. These peaks agree with the Patterson projections.

---

* Only one, inasmuch as the second sign that is determined for the projections by the rules indicated must be retained for reflections with $k = 2n + 1$, and this is what was done later.

** This separation of the reflections with $\underline{k}$ even into two subgroups was the result of the very weak participation in the statistics of reflections with $k = 2n + 1$.

*** Of the total number of $0kl$ (201) reflections, 125 have $k = 2n$, and 76 have $k = 2n + 1$. Thus, the density of reflections with $k = 2n + 1$ is rather considerable in the electron density synthesis, but the reliability of their sign determination is small.

The xy projection could not be used for comparison because the c axis is too long — 14.28 A. The shortest axis is b, 6.70 A, and corresponds to the xz projection, but comparison with this plane is excluded [3] because it coincides with the only symmetry plane of a monoclinic group (glide). We had to turn to the method of Harker-Kasper inequalities, which justified itself during the deciphering of xonotlite-wollastonite structures.* Reducing the F^2_{h0l} to an absolute scale was done according to Vainshtein's method [13], by applying his tables for the case when $B \leq 1$ (which is characteristic for silicates). $B = 0.60$ was obtained by Wilson's method [14].

A supporting group of 84 stronger (in unitary amplitude U_{h0l}) reflections (approximately 30% of the total of 288) was analyzed by inequalities. Among these were 7 with $U > 0.5$, 24 with $U > 0.4$, and 52 with $U > 0.3$. The inequalities produced 10 strong relationships and 5 weaker ones:

$$1)\ S\,(0.0.18) = S\,(4.0.10) \cdot S\,(4.0.8)\ (a = ky)$$
$$2)\ S\,(0.0.18) = S\,(8.0.\bar{8})\ \cdot S\,(8.0.10)\ (a = n\gamma)$$
$$3)\ S\,(12.0.9) = S\,(0.0.18) \cdot S\,(12.0.\bar{9})\ (\psi = as)$$
$$4)\ S\,(2.0.\overline{23}) = S\,(2.0.13) \cdot S\,(4.0.\overline{10})\ (i = bk)$$
$$5)\ S\,(10.0.17) = S\,(2.0.13) \cdot S\,(8.0.4)\ (u = bm)$$
$$6)\ S\,(14.0.4) = S\,(12.0.\bar{9}) \cdot S\,(2.0.13)\ (\varphi = sb)$$
$$7)\ S\,(12.0.9) = S\,(4.0.17) \cdot S\,(8.0.\bar{8})\ (\psi = cn)$$
$$8)\ S\,(10.0.17) = S\,(4.0.\overline{10}) \cdot S\,(14 \cdot 0.7)\ (u = kx)$$
$$9)\ S\,(20.0.\bar{5}) = S\,(14.0.7) \cdot S\,(6.0.\overline{12})\ (\delta = xl)$$
$$10)\ S\,(14.0.\overline{14}) = S\,(0.0.13) \cdot S\,(14.0.\bar{1})\ (t = dw)$$

$$1)\ S\,(12.0.\overline{18}) = S\,(0.0.18) \cdot S\,(12.0.0)\ (q = a\sigma)$$
$$2)\ S\,(14.0.\overline{14}) = S\,(0.0.18) \cdot S\,(14.0.4)\ (t = a\varphi)$$
$$3)\ S\,(2.0.13) = S\,(6.0.3) \cdot S\,(4.0.\overline{10})\ (b = \gamma k)$$
$$4)\ S\,(10.0.\overline{13}) = S\,(2.0.13) \cdot S\,(12.0.0)\ (v = b\sigma)$$
$$5)\ S\,(12.0.12) = S\,(8.0.4)\ \cdot S\,(4.0.8)\ (r = my)$$

In the second stage of the analysis the number of letters was reduced to 4: a, b, c, l, with the aid of statistical equalities; but one of them was excluded as a result of sign determinations from $h1l$ reflections as a consequence of the relationships

$$S\,(7.1.\bar{7}) \cdot S\,(2.0.\overline{23}) \quad \left.\right\} b$$
$$S\,(5.1.16) = S\,(7.1.\bar{7}) \cdot S\,(12.0.9) \quad \left.\right\} bl$$
$$S\,(8.0.10) \cdot S\,(3.1.\bar{6}) \quad \left.\right\} bcl$$
$$S\,(2.0.22) \cdot S\,(3.1.\bar{6}) \quad \left.\right\} bc$$
$$S\,(5.1.\bar{5}) = S\,(4.0.17) \cdot S\,(9.1.12) \quad \left.\right\} c$$
$$S\,(14.0.7) \cdot S\,(9.1.12) \quad \left.\right\} cl$$

. . . .

There turned out to be 16 equalities like these which confirmed that l = +, and thus for the four types of $h0l$ ** reflections we obtained

$$\begin{array}{ll} \text{1. } h = 4n,\ l = 2n, & \text{II. } h = 4n,\ l = 2n+1 \\ \quad +;\ c & \qquad a;\ ac \end{array}$$

$$\begin{array}{ll} \text{III. } h = 4n+2,\ l = 2n, & \text{IV. } h = 4n+2,\ l = 2n+1 \\ \qquad b;\ bc & \qquad ab;\ abc \end{array}$$

---

* Especially for 3×2 projections of datolite, herderite, gadolinite (all with the same group $P2_1/a$ [11]).

** In this analysis all h were even as a result of the halving of the a parameter of the cell due to the glide plane, which coincides with xz with a translational component $\parallel$ x.

In a two-dimensional projection one can manage with two letters; they must however agree with the two letters used for another projection. This was easy to do by using the $00l$ reflections common to xz and yz; and the plus sign was required for a, b was chosen as the second plus, and c must be a minus, otherwise all the amplitudes would become positive.

To check the results the signs of the reflections of the supporting groups, $0kl$, and $2kl$, and $4kl$ were used (to find the signs of $2kl$ and $4kl$ a combination of the comparison and statistical methods was applied). In the parallel determination of the signs of the $h0l$ reflections by the statistical method, using a new group of reflections, no contradiction was encountered with the results obtained earlier; this was taken as a confirmation of the accuracy of the signs established from the supporting group. The signs of all the remaining $h0l$ reflections were determined statistically from the 85 reflections of the $h0l$ supporting groups. Of the total of 288, 44 remained unestablished, i.e., 15.5%.

The projection of the electron density σ (xz) was constructed from 244 reflections. All the heavier atoms appeared on it, but peaks that would correspond in height to the O atoms turned out to be much greater in number than would be expected from the formula of spurrite.

The conventional (modulated [14]) Patterson projection constructed from the $h1l$ reflections enabled us to draw a very important conclusion concerning the location of the origin of the unit cell. The origin of the cell that we established (due to the accepted pulses for a and b) for the center of symmetry of the xz projection does not agree with that of the three-dimensional cell, and only the screw axis 2_1 passes through it, i.e., b must necessarily be ($-$); this was confirmed by further calculations when establishing the signs of the $h1l$ and $h2l$ reflections (their total number is 532 and 489).

In the corresponding $h1l$ supporting group 114 reflections (approximately 20%) were included, two of them with $U > 0.5$, 8 with $U > 0.4$, and 56 with $U > 0.3$. This group was strengthened by the previously determined signs of the $h0l$ supporting group.

The signs of the stronger reflections were determined with the aid of inequalities, the others by the statistical method.

In locating the signs of the $h2l$ supporting group (489 reflections) and using the $h0l$ and $h1l$ supporting groups we thought it feasible to proceed with the statistical method exclusively. All the others, i.e., those not included in the supporting groups, were established statistically with the following result: 102 $h1l$ and 92 $h2l$ reflections were not determined (19.2 and 18.8%). The magnitudes of F_{h1l} and F_{h2l} with signs were useful for the synthesis of the conventional (modulated) projections of the electron density: σ_1^{sin} from 176 reflections with $h = 2n$, σ_1^{cos} from 247 reflections with $h = 2n + 1$.

The locations of all the Ca^{2+} and Si^{4+} cations and also of 10 O atoms were determined from four projections of the electron density using $h0l$, $h1l$, $h2l$ (the three latter ones modulated). The conventional projections played a substantial part in solving the overlapping which occurred.

With the aid of the coordinates established by the syntheses described above, the signs of all the $h0l$ amplitudes were computed, and a new projection σ (xz) was constructed from which all the oxygen peaks seemingly superfluous in the first projection disappeared, and a peak emerged which corresponded to the light C.

The signs of 244 out of 288 $h0l$ reflections were determined by direct methods. As a result of recomputing according to the new electron-density synthesis, six reflections changed signs: among them one pinacoid remote from the supporting group (0.0.12), the other five were with small F.

Out of the 44 reflections not determined by direct methods 35 reflections obtained signs at this stage. For 20 reflections the calculation from coordinates led to too small amplitudes; therefore, it was thought impossible to include the corresponding signs in the final synthesis, which was actually constructed from 259 F_{h0l}.

For the final establishment of the x and z coordinates and simultaneously for a less exact determination of the y coordinates, we reverted to the combined modulated zone projections $S_{2/3}^{[-3/4, 0]}$ and $S_{2/3}^{[0, 3/4]}$ [15], which are easily constructed by combining with definite weight factors the zero and first rotation layer lines on b, and are convenient for the elucidation of atomic positions with coordinates $y = \frac{1}{4}$ $(\frac{3}{4})$ (as a result of dismemberment of the overlappings which occur in the xz projection). The corresponding formulas are

$$S^{[0, \, ^3/_4]}_{^1/_4} = \Sigma h \Sigma l \left[\left(F_{h0l} - \frac{F_{h1l}}{1.25} \right) \cos 2\pi \, (hx + lz) - \frac{F_{h1l}}{1.25} \sin 2\pi \, (hx + lz) \right]$$

The coordinates of the 19 basic atoms of spurrite (57 parameters) were obtained as a result of the analysis, and the comparison of all the projections are given in the table (the upper figure of each double line). The x and z coordinates established on the basis of strong relationships and from a large number of syntheses are more exact, and y is less so. The discrepancy coefficient R from 263 nonzero h0l reflections (up to sin λ = = 1.00) at this stage of the analysis was 25.0%, with very satisfactory distances between all the atoms (indicated in the closing part of the article).

The Coordinates of the Basic Atoms of the Spurrite Structure
(in hundredths of the a, b, c axes)

Atoms	x	y	z	Anions	x	y	z
Ca I	14.0	62.0	7.0	O III	0.4	30.0	8.3
	14.0	66.1	7.2		0.8	33.4	8.3
Ca II	3.2	25.0	61.3	O IV	25.3	32.0	10.6
	3.3	24.0	61.2		26.3	33.3	10.6
Ca III	24.2	5.0	45.6	O V	88.3	16.0	46.0
	23.9	6.1	45.5		88.3	15.3	46.8
Ca IV	4.7	99.0	83.3	O VI	89.4	44.0	33.5
	4.7	99.0	83.0		89.0	42.7	33.5
Ca V	13.4	53.0	82.9	O VII	2.1	8.0	33.1
	13.3	50.7	82.6		2.4	7.2	33.0
Si I	13.3	20.0	5.4	O VIII	10.2	30.0	46.9
	13.5	22.1	5.6		9.8	34.0	46.8
Si II	2.5	75.0	60.4	O IX	30.5	75.0	23.3
	2.5	75.0	60.4		28.3	69.6	23.5
C	19.7	75.0	27.1	O X	6.9	75.0	22.9
	18.5	69.5	28.5		7.3	68.5	23.0
O I	12.5	23.0	93.5	O XI	21.0	75.0	37.3
	12.4	23.7	93.8		21.1	71.2	37.2
O II	13.7	97.0	6.2				
	14.1	99.5	9.8				

Just then we received J. V. Smith, I. L. Karle, and H. Hauptman's manuscript in which the structure of the mineral was deciphered. This study of spurrite, begun already in 1953, brought results only by the beginning of 1960. Throughout these long years the authors of this famous method [16], Hauptman and Karle, succeeded in determining 39 signs for h0l reflections and 28 for 0kl, which should be compared with the 244 signs determined by the Moscow method. Nevertheless, the rough coordinates obtained by the American authors on the basis of corresponding syntheses were sufficient for further refinement to a conclusive stage — the derivation of the most probable values by the method of least squares. The final discrepancy coefficient (for nonzero reflections) is 19.0%. (The original article gave 13%, but the authors had twice as few reflections as we had, and we thought it to be essential to recompute the coefficient.) The coordinates of the American authors are given under our coordinates in each double line of the table. The difference in the magnitude of x and z is quite insignificant; it is greater in the y coordinates, which we have not yet refined. The latter, in view of the exceptional coincidence of results, has now become superfluous, and thus we give only the interatomic distances, and also a description of the structure of spurrite, which has not been sufficiently defined from the crystallochemical point of view in the article of the American authors.

In the structure of spurrite there is a complete absence of that clarity and legibility which characterize the diorthosilicates with [Si$_2$O$_7$] groups and Ca and Na cations (and also Zr and Ti which replace the latter in zircono and titanosilicates) and due to which it was possible to speak of the "Second Chapter of the crystal chemistry of silicates" [17]. Apparently the connection between the large Ca and Na cations in uniform octahedra and [Si$_2$O$_7$] diorthogroups is reciprocal, and in the orthogroups is reciprocal, and in the orthosilicates, such as spurrite, firstly, several kinds of Ca atoms are found to be in a different coordination, and secondly, the octahedra which are still preserved around a part of the Ca atoms are substantially distorted. This occurs in the high temperature structure of β Ca$_2$SiO$_4$, in afwillite Ca$_3$[SiO$_3$OH]$_2$·H$_2$O [18], in bultfonteinite Ca$_4$Si$_2$O$_{10}$F$_2$H$_6$ [19], and is very clearly apparent in spurrite, especially when compared with the tilleyite which has a similar formula Ca$_5$[Si$_2$O$_7$] (CO$_3$)$_2$.

The simplest way of describing the structure of spurrite is to derive it from the reaction according to which it is synthesized,

$$2Ca_2SiO_4 + CaCO_3 = Ca_5 [SiO_4]_2 CO_3,$$

and actually to see in the structure (Fig. 1) along the long axis c = 14.20 A layers from the orthosilicate α Ca$_2$SiO$_4$ of the olivine type alternating with the layers of CaCO$_3$ structure, but of aragonite and not of calcite. The Ca in the olivine-like layers have the coordination number 6 (octahedral), for the Ca in the aragonite layers it must be 12 (perovskite type CaTiO$_3$), but then it becomes distorted (as in aragonite itself) down to 9 [20]. In the olivine-like layers between two rows of Ca octahedra is squeezed the characteristic orthosilicate motif composed of Si tetrahedra stretched into a line, which in pairs look alternately up and

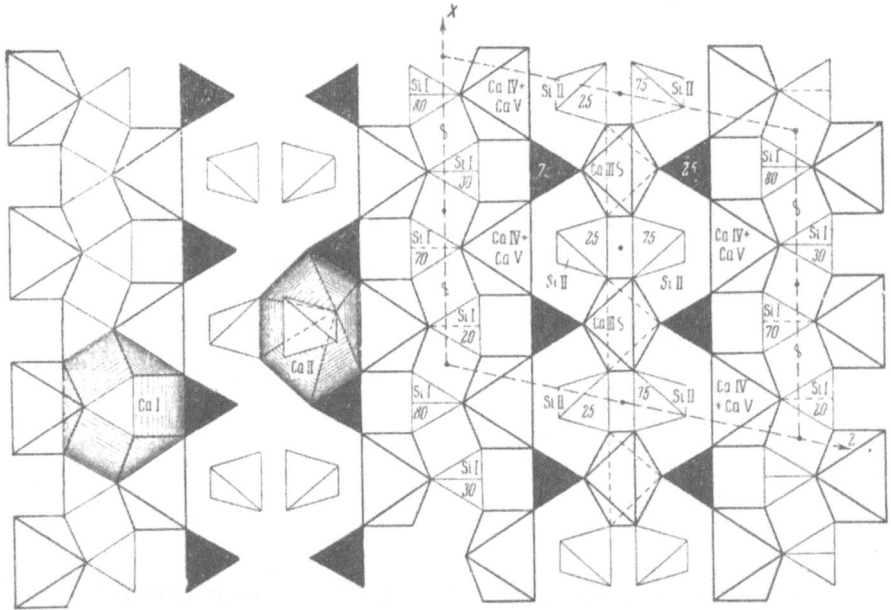

Fig. 1. Diagram of the structure of spurrite, xz projection. A true monoclinic cell is indicated in the background of the orthorhombic motif. Inside the CO$_3$ triangles and also in each of the two kinds of SiO$_4$ tetrahedra the height of the central atom is indicated (C or Si) by the y coordinate. Ca III octahedra are situated along the height of the cell in pairs one above the other with the coordinates y ± 0.05, 0.50 ± 0.05 and the Ca IV and Ca V octahedra also overlap each other with y coordinates near to 0 and to 0.50. The Ca I and Ca II polyhedra are not shown inside the monoclinic cell indicated. They are distributed above and below the two kinds (Si I and Si II) of tetrahedra and are illustrated (one of each kind) on the left of the diagram.

down from the b axis (see the structures of olivine and kyanite and their detailed description in [20]). This motif is isolated in Fig. 1 by a dashed line passing through it (x). The height of the b layer is 6.70 A and has two

Ca octahedra (Ca IV and Ca V) superimposed one upon another at a common face, i.e., according to the law of hexagonal closest packing, as is characteristic for olivine, but on the outside of the row (on the other side of the Si tetrahedra) the octahedra are distorted; the outermost is the sixth O atom and it lies on the level of the Ca atom.

As has been shown above, there is an "aragonite" layer between two olivine layers (α Ca_2SiO_4) in which similar situations occur as in aragonite; the polyhedra around Ca II are a flat hexagon and above it the edge of a Si tetrahedron is inserted between (each) two Ca II polyhedra, and under the latter there is only one vertex of a Si tetrahedron for each (thus, as in aragonite, Ca II has the coordination number 9).

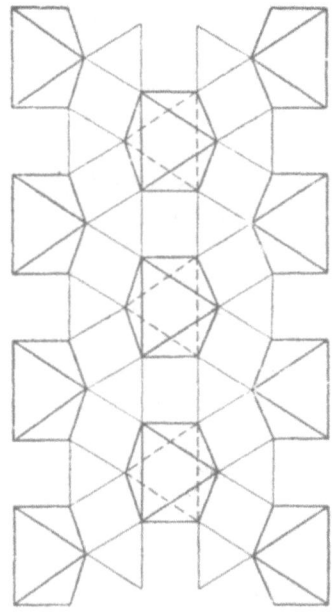

The Si nona vertica (Ca II) at two heights in this layer = 0.25 b and 0.75 b and the CO_3 groups are arranged between them at the same heights. However, exactly in the center of the aragonite layer, Ca atoms of the third kind (Ca III) are enclosed and form vertical b axis octahedral columns joined by common horizontal edges. If the CO_3 and SiO_4 groups are joined into an olivine motif similar to the one mentioned above, then, if previously there was a series of Si tetrahedra between two Ca octahedral rows of the olivine motif, then now two tetrahedral rows are removed from the olivine motif (CO_3 triangles are substituted for half of them) and substituted by one interstitial row of Ca III octahedra.

It is important to note that also in the layer which we called olivine under each Si tetrahedron is inserted a Ca polyhedron of the distorted aragonite type (Ca I) with a coordination number of 10 composed of six O atoms in a plane perpendicular to the b axis plus three O's from the bases of the overlying Si tetrahedra plus the apex of the underlying Si tetrahedron. Through the center of each of the separate layers — both the olivine and the aragonite — 2_1 screw axes parallel to b pass and the center of symmetry is between them. Both symmetry elements connect similar polyhedra from two heights (~ 0, $\sim {}^1/_2$, ~ 0.25, ~ 0.75), determined by the half-constant b/2. On one of the 2_1 axes the Ca III octahedra are strung out in a solid column. It is important that the particular (001) plane in which the 2_1 axes and centers of symmetry are concentrated in the center of the aragonite layer coincide with a pseudoglide plane with a translational component parallel to the b axis, whereas for the analogous plane (2_1 axes) in the center of the olivine layer the pseudoglide plane has a translational component parallel to the a axis. It is this last circumstance which shifts the origin that we have chosen to the center of symmetry of the olivine motif, and, after each (translational) step normal to the x axis, this origin is displaced by a quarter of a along the x axis, so that the cell becomes monoclinic with a completely "rational" angle $\beta = $ arc sin ${}^1/_4 \cdot$ a : c = = arc sin ${}^1/_4 \cdot 10.49 : 14.16 = 100°$ 40' (100° 19', in fact). The coupled Ca octahedra III, IV, and V determine the pseudo constant b/2.

Fig. 2. Mechanism of poly-synthetic twinning in spurrite. The Ca IV octahedra lose their one-sidedness on the twin plane and become regular. On both sides of the plane regular octahedra are located the one-sided Ca IV octahedra characteristic of spurrite, which corre-spond to the different elements of the twin.

The intrinsically orthorhombic design of the monoclinic structure (Fig. 1) was a great hindrance to Karle and Hauptman in their calculations, because the majority of the heavy atoms were concentrated in (401) planes, but this had little effect on the method of solution discussed here.

The planes parallel to (001) passing through the center of the aragonite layer along the columns of Ca octahedra, or through the centers of the Ca octahedra along the boundaries of the olivine layer, have good cleavage in some places, and the Si tetrahedra and the C triangles are distinctly separated, so that only the Ca — O bonds tear. The twinning planes are parallel to the same (001) plane. The latter arise when the Ca III octahedra with an already regular distribution of five vertices have also a "correct" sixth vertex. This circum-stance makes the lattice of these octahedra into b glide plane and locates in projection all its material content perfectly symmetrically in relation to this plane (Fig. 2).*

---

* The explanation for this cleavage given in the articles of the American authors seems to us to be highly formal and artificial.

From Fig. 1 we see that the characteristics of the (101) plane are similar, with the only difference that in it we deal with the overlapping of three kinds of columns of Ca octahedra, which makes this plane less expressed as twins, and also makes it a secondary cleavage plane. It is self-evident that the parallel distribution of all the CO_3 triangles makes spurrite sharply negative with its acute bisectrix parallel to the b axis, analogous to calcite.

Interatomic Distances in Spurrite

So far we have been describing the structure of spurrite on the basis of the ideal diagram of Fig. 1. Its detailed study, in particular the determination of all the interatomic distances, violates the perfection of this monoclinic structure. We shall pursue our study from the refined structure, noting the different results of the unrefined variant only in special instances. The distances Si I—O and Si II—O are 1.63 and 1.64 A, which is above the standard of approximately 1.60 A, as was noted by the American authors. We obtained more exact figures, 1.607 and 1.602 A, and likewise the edges of our Si tetrahedra are shorter (2.62 as against 2.65 and 2.67 A). The C—O distances = 1.21-1.36 A, with an average of 1.28 A, are better in the articles of the American authors than in ours (1.38 A). The Ca—O distances are very irregular, as is characteristic for all the Ca (and Na) silicates in which the similarity of the cation-anion distances is not fixed by symmetry (calcite, perovskite). When the sum of the ionic radii is approximately 2.4 A, spurrite has such low values as 2.27; 2.28; 2.30 A (American data); 2.21; 2.22, 2.29 (our data) and, furthermore, they increase at random, so that the exact coordination number is not always reliable. If the Ca—O distances were restricted to the smaller 2.8 A ones, than the Ca I and Ca II atoms would have a coordination number of 8 and 8, respectively, instead of 9 and 10. In the pseudo octahedra there will be 8 neighbors for Ca III instead of six, and for Ca IV and Ca V - seven instead of six. The Ca—O average distances calculated from the coordinates are amazingly close in both our and the American work: 2.47 and 2.48 A.

LITERATURE CITED

1. J. V. Smith, "Crystal data for spurrite $5CaO \cdot 2SiO_2 \cdot CO_2$," Acta Cryst. 8, 290 (1955).

2. J. V. Smith, "The crystal structure of tilleyite," Acta Cryst. 6, 9 (1953).

3. R. F. Smirnova , et al., "Crystal structure of cuspidine," Zap. Vses. Mineralog. o-va 84, 159 (1955).

4. J. D. McConnel, "A chemical optical and x-ray study of scawtite from Ballycraigy," Am. Mineralogist 40, 510 (1955).

5. D. Harker and J. S. Kasper, "Phases of Fourier coefficients directly from crystal diffraction data," Acta Cryst. 1, 70 (1948).

6. Kh. S. Mamedov and N. V. Belov, "Crystal structure of wollastonite," Doklady Akad. Nauk SSSR 107, 463 (1956).

7. Kh. S. Mamedov and N. V. Belov, "Crystal structure of minerals of the wollastonite group," I. Xonotlite, Zap. Vses. Mineralog. o-va 85, 13 (1957).

8. I. M. Rumanova, "Sign determination of the 'supporting' structure amplitudes by statistical equalities," Tr. Inst. Kristallografiya Akad. Nauk SSSR 10, 57 (1954).

9. E. G. Fesenko, I. M. Rumanova, and N. V. Belov, "Crystal structure of zoisite," Kristallografiya 1, 165 (1956).

10. N. V. Belov, "Classroom method of deriving symmetry space groups," Tr. Inst. Kristallografiya, Akad. Nauk SSSR 6, 25 (1951).

11. P. V. Pavlov and N. V. Belov, "Crystal structure determination of herderite, datolite, and gadolinite by direct methods," Kristallografiya 4, 324 (1959). [Soviet Physics — Crystallography, Vol. 4, p. 300].

12. S. V. Borisov, V. P. Golovachev, and N. V. Belov, "Concerning the arbitrarily assigned signs in direct methods of deciphering crystal structures," Kristallografiya 3, 269 (1958). [Soviet Physics — Crystallography, Vol. 3, p. 274].

13. B. K. Vainshtein, "Quantitative relationships in Fourier series of the electron density of crystals," Zhur. Éksp. i Teoret. Fiz. 27, 44 (1954).

14. A. J. C. Wilson, "Determination of absolute from relative x-ray intensity data," Nature 150, 151 (1942).

15. I. M. Rumanova, "Modulated zonal projections of electron density in structure analysis," Kristallografiya 4, 143 (1959). [Soviet Physics — Crystallography, Vol. 4, p. 127].

16. H. Hauptman and J. Karle, "Solution of the phase problem. I. The centrosymmetrical crystal," ACA Monograph (1953), No. 3.

17. N. V. Belov, "The second chapter of the crystal chemistry of silicates," Zhur. Strukt. Khim. $\underline{1}$, 1 (1960).

18. H. D. Megaw, "The structure of afwillite $Ca_3[SiO_3OH]_2 \cdot 2H_2O$," Acta Cryst. $\underline{5}$, 477 (1952).

19. E. I. McIver and H. D. Megaw, "The structure of bultfonteinite $Ca_4Si_2O_{10}F_2H_6$ and its thermal dehydration products," Symposium über Silikate mit ein- und zweiwertigen Kationen (Berlin, 1960).

20. N. V. Belov, The Structure of Ionic Crystals and Metallic Phases [in Russian] (Moscow, 1947).

Reprinted from Soviet Physics—Doklady, Vol. 5, pp. 1141-1144, May-June, 1961

THE CRYSTAL STRUCTURE OF HURLBUTITE

V. V. Bakakin and Academician N. V. Belov

Translated from Doklady Akademii Nauk SSSR, Vol. 135, No. 3, pp. 587-590,
November, 1960
Original article submitted August 18, 1960

In previous papers [1, 2] some of our results on the x-ray investigation of hurlbutite, $CaBe_2P_2O_8$, have been presented. These discussed the general shape in relation to its similarities to and differences from the similar data for danburite, $CaB_2Si_2O_8$. The absence of exact values for the x coordinates did not allow a complete solution of the hurlbutite structure. Below we fill this gap and, at the same time, we are able to demonstrate a further case of the structural analogy between certain phosphates and silicates.

TABLE 1

Coordinates of the Basic Atoms in the Structure of Hurlbutite

Atom	x	y	z	Atom	x	y	z
Ca	0.386	0.085	0.753	O_3	0.121	0.367	0.438
Be_1	0.059	0.196	0.435	O_4	0.126	0.364	0.055
Be_2	0.265	0.421	0.933	O_5	0.412	0.308	0.565
P_1	0.264	0.418	0.560	O_6	0.415	0.309	0.931
P_2	0.059	0.197	0.060	O_7	0.006	0.150	0.247
O_1	0.188	0.083	0.508	O_8	0.184	0.421	0.745
O_2	0.189	0.085	0.993				

Hurlbutite is $CaBe_2P_2O_8$; the parameters of its monoclinic (pseudorhombic) cells are: $a - 8.29$ A, $b = 8.80$ A, $c = 7.81$ A, $\beta = 90°$; space group $P2_1/a$; $Z = 4CaBe_2P_2O_8$. The basic experimental material consisted of 184 F_{hk0}, 171 F_{h0l}, and F_{0kl} (KFOR camera, Mo-radiation).

In [2], by comparing the structures of danburite and hurlbutite to analyze the latter, the projection yz was reduced to an error factor $R_{0kl} = 17.5\%$ for all nonzero reflections. It could be reduced further to 14.1% for all nonzero reflections, and to 18.3% with the inclusion of 88 zero F_{exp} (to $\sin \vartheta/\lambda = 1.05$ A^{-1}).

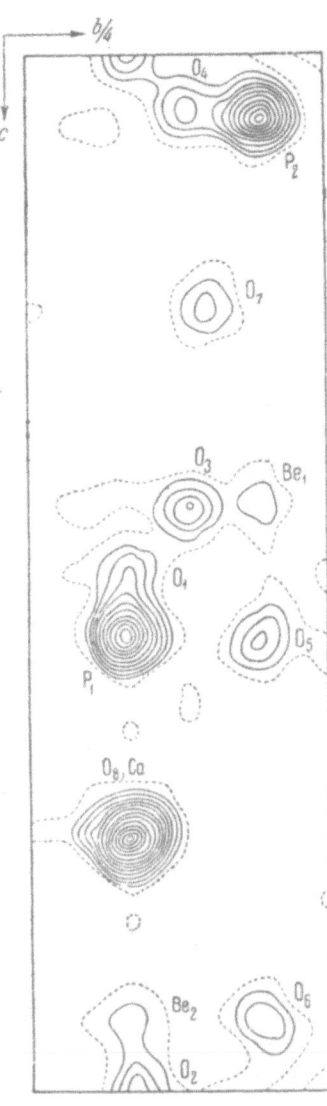

Fig. 1. Hurlbutite. Projection
of the electron density on yz
in contours of equal density.

TABLE 2

Interatomic Distances (in A), Angles, and the Valency Balance in
Hurlbutite

P_I-tetrahedra		P_{II}-tetrahedra		Be_I-tetrahedra		Be_{II}-tetrahedra	
P_I—O_I	1.60	P_{II}—O_{II}	1.55	Be_I—O_I	1.57	Be_{II}—O_{II}	1.60
P_I—O_{III}	1.58	P_{II}—O_{IV}	1.57	Be_I—O_{III}	1.59	Be_{II}—O_{IV}	1.58
P_I—O_V	1.56	P_{II}—O_{VI}	1.56	Be_I—O_V	1.58_5	Be_{II}—O_{VI}	1.59
P_I—O_{VIII}	1.59	P_{II}—O_{VII}	1.58	Be_I—O_{VII}	1.58_5	Be_{II}—O_{VIII}	1.61
O_I—O_{III}	2.51	O_{II}—O_{IV}	2.56	O_I—O_{III}	2.62	O_{II}—O_{IV}	2.51
O_I—O_V	2.62	O_{II}—O_{VI}	2.50	O_I—O_V	2.52	O_{II}—O_{VI}	2.64
O_I—O_{VIII}	2.66	O_{II}—O_{VII}	2.56	O_I—O_{VII}	2.60	O_{II}—O_{VIII}	2.72
O_{III}—O_V	2.66	O_{IV}—O_{VI}	2.51	O_{III}—O_V	2.52	O_{IV}—O_{VI}	2.63
O_{III}—O_{VIII}	2.50	O_{IV}—O_{VII}	2.60	O_{III}—O_{VII}	2.60	O_{IV}—O_{VIII}	2.52
O_V—O_{VIII}	2.54	O_{VI}—O_{VII}	2.61	O_V—O_{VII}	2.63	O_{VI}—O_{VIII}	2.60

Ca-polyhedra		Bond angles P — O — Be		Valency balance	
Ca—O_I	2.52	$\angle P_I$ —O_I —Be_I	129°	O_I	$=2+^1/_{28}$
Ca—O_{II}	2.49	$\angle P$ —O_{II} —Be_{II}	127°	O_{II}	$=2+^1/_{28}$
Ca—O_{III}	2.46	$\angle P_I$ —O_{III} —Be_I	121°	O_{III}	$=2+^1/_{28}$
Ca—O_{IV}	2.42	$\angle P_{II}$ —O_{IV} —Be_{II}	125°	O_{IV}	$=2+^1/_{28}$
Ca—O_V	2.43	$\angle P_I$ —O_V —Be_I	127°	O_V	$=2+^1/_{28}$
Ca—O_{VI}	2.46	$\angle P_{II}$ —O_{VI} —Be_{II}	128°	O_{VI}	$=2+^1/_{28}$
Ca—O_{VIII}	2.47	$\angle P_{II}$ —O_{VII} —Be_I	135°	O_{VII}	$=2-^1/_4$
		$\angle P_I$ —O_{VII} —Be_{II}	131°	O_{VIII}	$=2+^1/_{28}$
Ca—O'_{III}	3.16				
Ca—O'_{IV}	3.12	Mean	128°	Control	$^7/_{28}=^1/_4$

TABLE 3

	a, A	b, A	c, A	β	$a:b:c$	Space group
Danburite	8.01	8.75	7.71	$=90°$	$0.915:1:0.891$	$D_{2h}^{16}=Pnam$
Hurlbutite	8.29	8.80	7.81	$\approx90°$	$0.942:1:0.887$	$C_{2h}^5=P2_1/a$
Paracelsian	9.08	9.58	8.58	$\approx90°$	$0.947:1:0.895$	$C_{2h}^5 - P2_1/a$

This projection is reproduced in its final form in Fig. 1, which should be compared with Fig. 2b to demonstrate the possibility of distributing the "heavy" P and light Be atoms very exactly in the positions which are occupied in danburite by pairs of Si atoms (from Si_2O_7 diorthogroups) or by pairs of B atoms (from B_2O_7 groups).

The x coordinates of all atoms, which we had previously established only from crystal chemical considerations, are now made more exact by electron density projection on the xz plane, and by partial projection on the xy plane in which 10 out of the 13 atoms overlap in pairs. The values of the error coefficients are: R_{h0l} = 18.1% and R_{hk0} = 17.0% (to sin ϑ/λ = 1.05, F $\neq$ 0).

The final coordinates of the basic atoms are cited in Table 1. All the atoms occupy general positions, and hence hurlbutite structure is characterized by 39 independent parameters.

As was shown earlier [2], the structure of hurlbutite is similar to the structure of danburite, $CaB_2Si_2O_8$; both have a skeleton constructed from four-membered and stretched eight-membered centrosymmetric rings made of two sorts of tetrahedra — a skeleton which, apart from the orientation of the rings, is characteristic of the feldspars. In hurlbutite, the general arrangement of the atoms is the same as in danburite, but in the latter, each tetrahedron is coupled (with an apical O in common) to another of like kind (and the corresponding O atom appears in the plane of symmetry) and with three of the other sort, whereas in hurlbutite each PO_4-tetrahedron is joined to four

147

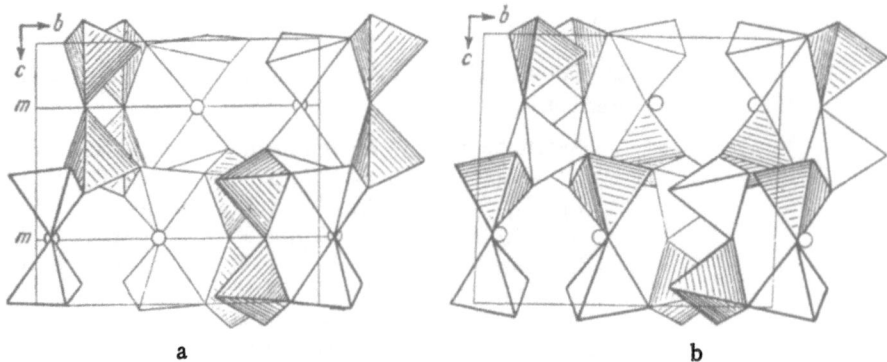

Fig. 2. Similar yz projections of the structure of: a) danburite, b) hurlbutite. The hatched tetrahedra in a are SiO_4 and in b are PO_4. There is a mirror plane only in the first structure, in which it passes through the central O in the $[Si_2O_7]$ and $[B_2O_7]$ groups.

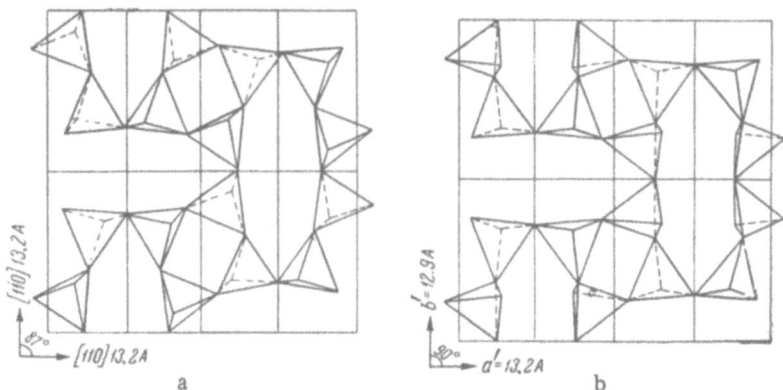

Fig. 3. Comparison of two pseudotetragonal projections: a) hurlbutite, b) anorthite with alternating octameric and tetrameric rings formed from tetrahedra. The tetrahedra appearing above the plane of the diagram are drawn with thickened lines; those appearing below the plane with broken lines.

BeO_4-tetrahedra, and vice versa (Figs. 2a and 2b). The Ca atoms are positioned in their characteristic seven-apexed coordination position (a trigonal prism plus half an octahedron) elongated along the c-axis at the center of the pseudoquadratic base xy of the cell.

The basic values which characterize the structure are given in Table 2. The P—O distances fall within the limits 1.55-1.60 A, and the Be—O distances = 1.57-1.61 A. The O—O edges in the PO_4- and BeO_4-tetrahedra are, respectively, 2.50-2.66 A and 2.51-2.72 A. The seven Ca—O distances are equal to 2.42-2.52 A (the two next nearest distances are greater than 3.1 A).

We have seen that the PO_4 and BeO_4 tetrahedra are nearly equal in size, and both types are shown to be intermediate between the SiO_4 and BO_4 sizes in danburite.

As in the case of danburite, attention is drawn to the closeness of all the Ca—O distances in the seven-apexed figure around Ca. In other mineral structures, the 6, 7, or 8 distances from the loose Ca cation to the nearest O are usually characterized by a large scatter.

As has been shown above, the structures of danburite $CaB_2Si_2O_8$ and hurlbutite are similar to the structure of the feldspars, in particular to anorthite $CaAl_2Si_2O_8$. In order to show the alternation in all three structures of eight-membered rings of tetrahedra with four-membered ones, it is necessary, for hurlbutite, to take the basocentric view of the pseudotetragonal projection (001), and then compare it with the pseudotetragonal projection of anorthite (Figs. 3a and 3b).

The structure of hurlbutite, $CaBe_2P_2O_8$, served as a key to the solution of the structure of paracelsian (δ-$BaAl_2Si_2O_8$). Although the unit cell of the latter is very similar to the cell of rhombic danburite, $CaB_2Si_2O_8$, paracelsian nevertheless appeared to be monoclinic and to be exactly analogous to monoclinic hurlbutite with a strict distribution of Si and Al in different positions, analogous to the positions of P and Be in hurlbutite.

A complete solution for paracelsian appears in [3].

LITERATURE CITED

1. V. V. Bakakin and N. V. Belov, Doklady Akad. Nauk SSSR 125, 383 (1959).

2. V. V. Bakakin and N. V. Belov, Doklady Akad. Nauk SSSR 129, 420 (1959).

3. V. V. Bakakin and N. V. Belov, Kristallografiya 5, No. 6 (1960) [in press].

Reprinted from Soviet Physics—Crystallography, Vol. 5, pp. 826-829, May-June, 1961

CRYSTAL STRUCTURE OF PARACELSIAN

V. V. Bakakin and N. V. Belov

Institute of Inorganic Chemistry, Siberian Division, Academy of Sciences of the USSR
Translated from Kristallografiya Vol. 5, No. 6, pp. 864-868, November-December, 1960
Original article submitted August 10, 1960

After we had established the structure of hurlbutite [1-3] we found on comparing the structure with those of danburite and the feldspars that one form of celsian, δ $BaAl_2Si_2O_8$ or paracelsian [4], is very similar not only to danburite $CaB_2Si_2O_8$, as has been pointed out [4-6], but also to hurlbutite $CaBe_2P_2O_8$.

	a, A	b, A	c, A	β	$a:b:c$	Space group
Danburite	8.01	8.75	7.71	$\beta = 90°$;	$0.915:1:0.881$;	$D_{2h}^{16} = Pnam$
Hurlbutite	8.29	8.80	7.81	$\beta \approx 90°$;	$0.942:1:0.887$;	$C_{2h}^5 = P2_1/a$
Paracelsian	9.08	9.58	8.58	$\beta \approx 90°$;	$0.947:1:0.895$;	$C_{2h}^5 = P2_1/a$

The hurlbutite-paracelsian pair has

$$a'':a' \approx b'':b' \approx c'':c' \approx 1.1.$$

A more detailed examination gave rise to the following conclusions.

Spencer [4] first described paracelsian in detail in 1942 for material from the Benallt Mine, Wales. Goniometric and optical studies showed that the crystals are strictly orthorhombic in form, but the optical measurements indicate a lower symmetry. Smith [5, 6] examined monocrystals from this same source in 1952-53; the powder, Weissenberg, and oscillation photographs showed that the mineral is monoclinic, but with a strictly pseudoorthorhombic unit cell: a = 9.08, b = 9.58, c = 8.58 A; $\beta = 90° \pm 10'$, Z = 4 $BaAl_2Si_2O_8$ ($\sigma = 3.31$). The absences imply space group P2$_1$/a, but some accidental absences cause the group to appear to be Pna2$_1$ or Pnam (unless the photographs are much overexposed). Smith found the structure by trial and error on the basis of the structure of danburite*; no distinction was made between Si and Al atoms, and it was assumed that they were randomly distributed over the sites for Si and B in danburite. The danburite-type orthorhombic form for the structure was refined by means of Fourier synthesis to give $R_{hk0} = 13\%$ and $R_{h0l} = 14\%$. No projections on the yz plane were calculated and 10 of the 13 atoms overlap in pairs in the xy projection, so no precise values could be given for the y coordinates of atoms associated with the mirror pseudoplane. The $\underline{x}$ and $\underline{z}$ of two pairs of (Si, Al) atoms and of the $\overline{O}$ of (Si, Al)$_2$O$_7$ remain of poor accuracy, because these overlap in the $\underline{xy}$ projection. Of the 39 parameters, 18 remain approximate. In some cases the $\underline{x}$ derived from the two projections do not agree very well. Smith thus gives a list of averaged values (for the danburite structure) together with the exact coordinates actually found; the analysis is performed with reference to that model, although the actual structure clearly deviates from it. He gives as probable reasons for the deviation that the specimens were twinned (at very small angles), that the atoms are slightly displaced for reasons of energy, and finally that a certain degree of order occurs in the array of Si and Al atoms, although he considers that perfect ordering could occur within the orthorhombic structure. His analysis implies that the two crystallographically distinct kinds of tetrahedron in the ideal model have Si:Al = 1:1, i.e. that a tetrahedron contains Si $\frac{1}{2}$ Al$\frac{1}{2}$ [the mean (Si, Al)-O distances are 1.68 and 1.71 A to a low accuracy, since the usual difference in the Si-O and Al-O distances is about 0.1 A].

---

*Spencer [4] stated that the cell parameters are similar to those of topaz, but Smith rejected the idea that there is any structural resemblance.

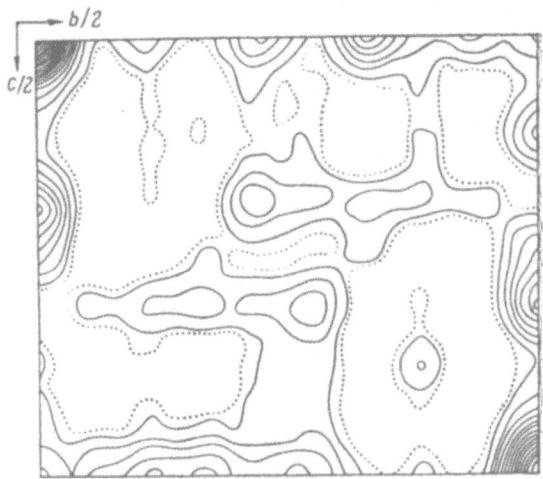

Fig. 1. Patterson projection on yz (one quadrant) for hurlbutite (paracelsian).

Smith would have that $Si_{\frac{1}{2}}Al_{\frac{1}{2}}$ expresses the crystallographic analysis; on this basis he compares the results with those for other aluminosilicates and proposes that the bond length is linearly related to the Si:Al ratio, with the consequence that he introduces new "standard" (and, in his view, exact) values for Si-O (1.60_5 A) and Al-O (1.78 A) for aluminosilicates; then the distances observed for paracelsian correspond exactly to $Si_{\frac{1}{2}}Al_{\frac{1}{2}}$.

A detailed study of Smith's work has shown us that the analysis has not been carried through properly. In any case, it is impossible to consider the $Si_{\frac{1}{2}}Al_{\frac{1}{2}}$ formula seriously when so many of the parameters are approximate, so we assumed that the distributions of the Si and Al remain unknown. Although Smith showed that paracelsian has no danburite Si_2O_7 and Al_2O_7 groups, he did not examine other possible ordered arrays for the Si and Al atoms. As soon as we had finished analyzing the structure of hurlbutite [3], we found that it was very probable that the cations in this mineral have the same distribution as those in paracelsian. The first indication of this is that the x-ray patterns of the two minerals have much in common. For example, paracelsian gives reflections of $0kl$ type with $k + l = 2n + 1$, which inevitably imply a plane of symmetry parallel to (100); these actually appear on overexposed films, whereas those of $00l$ type with $l = 2n + 1$ remain absent, although this feature is not an essential consequence of the space group. We have found the same effect [1], rather better developed, for hurlbutite, whose $0kl$ reflections with $k + l = 2n$ are rather stronger than the others of that type, whereas the $00l$ reflections with $l = 2n + 1$ do not appear at all. The reason for the weakened reflections is mainly that P and Be differ greatly in atomic number; this large difference enabled us to distinguish the atoms in terms of the "extra" reflections and to check the analysis on the electron-density patterns *. The effect has the same cause in the case of paracelsian, but is not nearly so strongly developed because there is only the minimum difference between Si and Al. We anticipated that we would be able to differentiate the Si and Al in terms of the volumes of their tetrahedra and that we would thereby be able to test for the hurlbutite pattern, so we analyzed Smith's results [6] with great care.

A great variety is seen in the complete set of (Si, Al)-O and O-O distances for the four kinds of tetrahedra. Some of this arises from the poor accuracy of 50% of the coordinates, so in our subsequent analysis we used only the most reliable ones (Smith's exact coordinates). We entirely neglected any bond that includes one of the four atoms that overlap in pairs on the xz projection; of the others we took account only of those whose lengths depend very little on the y coordinates, which remain undefined for most of the atoms. The four monoclinic tetrahedra can be split into two hurlbutite pairs, which differ on average by 0.12 A for the O-O distances and by 0.07 A for the (Si, Al)-O ones. The most reliable distances (in Smith's symbols) are 1.64 and 1.71 for $(Si, Al)_2-O_5$, 1.66 and 1.73 for $(Si, Al)_2-O_2$, 2.62 and 2.75 for O_2-O_5, etc. These results show that the Si actually is segregated from the Al in paracelsian; the two form an ordered array exactly as do P and Be in hurlbutite. We have calculated R for the ordered structure and for the structure with Si and Al disordered; the first gives a value better by 0.5% (by 3% of R itself). The Si and Al differ very little in scattering power, and Ba is very heavy, so this improvement is quite considerable.

The errors in the coordinates tended to obscure this ordered array of Si and Al atoms, so we have sought to revise the coordinates by means of the analogy with hurlbutite. The corrected coordinates, while not of very special precision, correspond better to the real structure than do ones averaged for overlapping atoms, the more so since many of the corrections do not exceed the error of experiment.

---

*The Patterson diagram represents the intensities most fully. Figure 1 shows the yz pattern for hurlbutite, which, if it were exactly analogous to danburite, would give a centered yz projection (in accordance with the $\underline{n}$ in Pnam), in which case the peak in the right lower corner of the quadrant would be exactly the one at the top left corner; in fact, it is some 30% weaker. The same is true for other peaks related by the pseudocenter at the middle of the quadrant; the effect is very marked for most such pairs, but not for all, the controlling factor being the relative contribution from P + Be to each such peak.

Coordinates of the Basal Atoms in Paracelsian (and Hurlbutite) ***

	x	y	z	x	y	z	
Ba	0.397 —	0.088 —	0.751 —	0.386	0.085	0.753	Ca
Si_I $(Si, Al)_I$	0.067 (0.065*)	0.196 —	0.435 (0.440**)	0.059	0.196	0.435	Be_I
Si_{II} $(Si, Al)_{II}$	0.270 —	0.417 —	0.938 —	0.265	0.421	0.933	Be_{II}
Al_I $(Si, Al)_{II}$	0.274 —	0.419 (0.417*)	0.571 —	0.264	0.418	0.560	P_I
Al_{II} $(Si, Al)_I$	0.064 (0.065*)	0.196 —	0.067 (0.069**)	0.059	0.197	0.060	P_{II}
O_I } (O_I)	0.194 (0.195*)	0.088 —	0.501 —	0.188	0.083	0.508	O_I
O_{II}	0.195 —	0.083 (0.088*)	0.003 —	0.189	0.085	0.993	O_{II}
O_{III} } (O_{II})	0.130 —	0.360 (0.362*)	0.454 —	0.121	0.367	0.438	O_{III}
O_{IV}	0.128 —	0.365 (0.362*)	0.046 —	0.126	0.364	0.055	O_{IV}
O_V } (O_{III})	0.428 (0.425*)	0.313 —	0.555 (0.560**)	0.412	0.308	0.565	O_V
O_{VI}	0.417 (0.425*)	0.316 (0.313*)	0.940 (0.931**)	0.415	0.309	0.931	O_{VI}
O_{VII} (O_{IV})	0.007 —	0.173 —	0.255 —	0.006	0.150	0.247	O_{VII}
O_{VIII} (O_V)	0.209 —	0.412 —	0.758 —	0.184	0.421	0.745	O_{VIII}

*Mean from the xy projection for two atoms related by pseudosymmetry.

**Mean for two atoms overlapping in the xz projection.

***The values in brackets are those from [6].

The Table gives the original coordinates and the revised ones, as well as the parameters for hurlbutite. Figures 2 and 3 show the main structural features of danburite and paracelsian.

The corrected coordinates give us revised distances as follows: Si-O and O-O in the SiO_4 tetrahedra, 1.65 and 2.70 A (average), respectively; Al-O and O-O in the AlO_4 tetrahedra, 1.72_5 and 2.82 A; Ba-O, seven values ranging from 2.73 to 2.83 A.

Thus paracelsian and hurlbutite have the same type of structure (Ba, Si, and Al replace Ca, Be, and P); the silicate has the same structure as the phosphate even though the cations are different.

The ordered array for the Si and Al agrees better than a random array with the experimental results; it is also in full agreement with current ideas on the distributions of the Si and Al in feldspars. For instance, in a recent [7] paper on celsian $BaAl_2Si_2O_8$ (the usual monoclinic form) it has been shown that the Si and Al alternate; each oxygen atom has one silicon atom and one aluminum atom as its neighbors. Reference is made there to the similar behavior of Si and Al in anorthite [8], and it is supposed (following Läwenstein [9]) that this alternation of Si and Al is electrostatically the most favorable if Al:Si = 1:1.

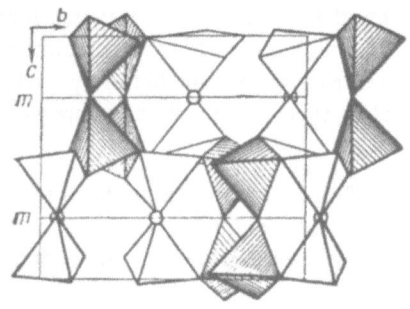

Fig. 2

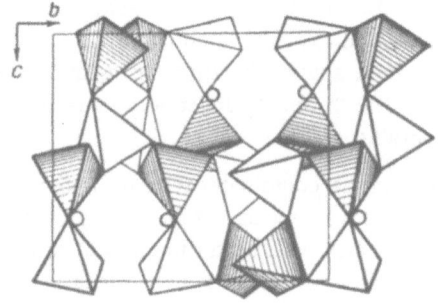

Fig. 3

The interatomic distances found for celsian (on average 1.64 and 2.67 A for Si-O and O-O in the Si tetrahedra, and 1.71_5 and 2.80 A for Al-O and O-O in the Al tetrahedra) agree well with the values we find for paracelsian; they do not agree with Smith's 'standard' values of 1.60_5 and 1.78 ± 0.02 A for Si-O and Al-O, respectively. However, in an attempt to avoid conflict with Smith's values, the explanation is offered for celsian that the Si and Al are not completely ordered, although no obvious causes for any effect on the degree of order are apparent. We consider that Smith's values for "pure" Si-O and Al-O bonds, which he discusses very one-sidedly in a later paper [10] on these distances, are unsound and must be revised.

Note. After this paper was written we saw the results of Smith et al. for spurrite [11]; the Si-O distances he gives there (1.626-1.637 Å) agree completely with our figure (1.64 Å). In that paper [11] doubt is cast on the earlier assumption that the Si-O distance is constant at 1.605 Å unless special causes are operative (we must remember here that this dogmatic statement of Smith's was made in 1953, when he was only 22 years old).

LITERATURE CITED

1. V. V. Bakakin and N. V. Belov, Doklady Akad. Nauk SSSR 125, 2, 383 (1959).
2. V. V. Bakakin, V. B. Kravchenko, and N. V. Belov, Doklady Akad. Nauk SSSR 129, 2, 420 (1959).
3. V. V. Bakakin and N. V. Belov, Doklady Akad. Nauk SSSR 135, 3 (1960). [See Soviet Physics — Doklady].
4. L. J. Spencer, Mineral. Mag. 26, 231, 1942.
5. J. V. Smith, Am. J. Sci. 513 (1952).
6. J. V. Smith, Acta Cryst. 6, 7, 513 (1953).
7. R. E. Newnham and H. D. Megaw, Acta Cryst. 13, 4, 303 (1960).
8. C. J. E. Kempster, Thesis, Cambridge University (1957).
9. W. Lawenstein, Am. Mineralogist 39, 92 (1954).
10. J. V. Smith, Acta Cryst. 7, 6-7, 479 (1953).
11. J V. Smith, I. L. Karle, H. Hauptman., and J. Karle, Acta Cryst. 13, 6, 454 (1960).

Reprinted from Soviet Physics—Crystallography, Vol. 6, pp. 685-693, May-June, 1962

THE CRYSTAL STRUCTURE OF RUBIDIUM DI (META) FLUORO-BERYLLATE (RbBe₂F₅) AND ITS RELATIONSHIP TO SILICATE SHEET STRUCTURES WITH [Si₂O₅] UNITS

V. V. Ilyukhin and N. V. Belov

Crystallography Institute, Academy of Sciences, USSR
Translated from Kristallografiya, Vol. 6, No. 6,
pp. 847-858, November-December, 1961
Original article submitted July 26, 1961

The crystal structure of Rb difluoroberyllate was determined from xz and yz Patterson diagrams. The method of integral characteristics played an important part in the determination. The structure suggests a "relieved" layer silicate. Brucite-type layers of Rb octahedra alternate with double flat $[Be_2F_5]_{\infty,\infty}$ networks made up from BeF_4 tetrahedra. The relationship was examined between this structure model and classical layer silicates, and also recently interpreted silicates containing no isolated packets, such as clay minerals, micas, and chlorites.

I

The chemical structures of ortho-, meta-, and di-meta- (layer) silicates have been very incompletely studied, and have always been complicated by the generally negligible solubility of silicates, the high melting points, and, to a large extent, the slow equilibrium of silicate systems. Since Goldschmidt's time, a number of investigators have tried to approach the chemistry or geochemistry of silicates through soluble "model silicates," the properties of which could be related to the corresponding silicates. The best known soluble model silicates have long been the fluoroberyllates [1-16].

Back in 1926-27, Goldschmidt [8, 9] pointed out the very close similarity between BeF_2 and SiO_2, in spite of the former's solubility and the latter's insolubility, and attributed this similarity to the closeness of ionic radii and polarization properties in fluorides and oxides, the comparative values being: Be^{2+}—0.34 A, Si^{4+}—0.39 A, F^{1-}—1.33 A, O^{2-}—1.33 A.

The halved charges on Be and F compared to Si and O weakens the bonding in fluoroberyllate lattices, leading to low refractive indices, low melting points, and good solubility and stability under ordinary conditions of temperature and pressure. The solubility permits comparatively easy preparation of single crystals of fluoroberyllates from aqueous solutions and melts.

Much of the data from fluoroberyllate systems [1-16] has already been used to resolve a number of silicate problems. It is also noteworthy that studies of the properties, composition and structure of fluoroberyllates are not only of theoretical interest from the model silicate viewpoint, but also have practical significance in the preparation of pure metals and their alloys. Here the role of fluoroberyllates is analogous to that of cryolite, Na_3AlF_6, in aluminum metallurgy, and of fluorotitanates in titanium chemistry.

Of the large number of fluoroberyllates (ortho-, meta-, and others), the least studied have been those with the "dimeta-" radical $[Be_2F_5]$, for which, in analogy with silicates, a layer structure would be expected. The simplest are compounds with the formula $MeBe_2F_5$, where Me represents an alkali metal. No success has been achieved in preparing Li and Na compounds [12], at least not in a form suitable for x-ray investigation. According to [3], Li, Na and K fluoroberyllates decompose before melting, and the only stable fluoroberyllates are those of Rb and Cs; these may be prepared from aqueous solution, and also by sintering the correct amounts of MeF and BeF_2 [3, 12]. Much of the work has been carried out by Novoselova and her colleagues in Moscow [3, 6, 10, 11, 14, 15, 16], and by Toropov and Grebenshchikov in Leningrad [2, 4, 5, 7, 12]. The latter kindly gave us a single-crystal specimen of $RbBe_2F_5$ for x-ray investigation.

This specimen resembled a mica, with almost perfect cleavage parallel to (001), plates of which on examination with the microscope clearly showed the conoscopic image of a biaxial crystal. The low indices of refraction, $n_{av} = 1/2(n_g + n_p) = 1.332$ [4], did not permit the determination (in immersion) of the very small birefringence. On the (001) plane, striations at 120° were clearly visible, corresponding to the (100) and (010) cleavages the crystals being morphologically pseudohexagonal. The specific gravity, by the density bottle method, was about

2.809. The slightly hygroscopic crystals underwent a polymorphic transformation on heating, $(\alpha \underset{301°}{\rightleftarrows} \beta \underset{80°}{\rightleftarrows} \gamma)$. The γ-modification, stable in ordinary conditions, showed a tendency towards polysynthetic twinning. The value of the optic axial angle was $2V > 40°$ (about 50°). The optic sign was as expected for the layer structure.

II

The platy, micalike form of the crystals, and the polysnythetic twinning were great hindrances to the determination of the lattice constants from oscillation photos, so that at first the crystal was considered monoclinic (pseudorhombohedral) crystal of class C2/m, with a = 7.99, b = 4.70, c = 6.12 A; $\beta \approx 90°$, and with two formula units of $RbBe_2F_5$. The pseudohexagonal form of the crystals tied up with the axial ratio a : b $\approx \sqrt{3}$ and with the centered (C) pseudo-orthohexagonal unit cell.

However, on the Weissenberg photos prepared for the zero-layer lines (Mo radiation), no lines (planes) of symmetry were observed for the side zones 0kl and hk0, indicating the lower symmetry of triclinic. C, the corresponding unit cell under this arrangement, with two possible space groups, C$\bar{1}$ and C1, has the constants: a = 7.98, b = 4.69, c = 6.12 A; $\alpha = 89°40'$; $\beta = 91°$; $\gamma = 90°27'$.

The transformation to the primitive unit cell is, for the axes, analogous to the transformation of the reference triad of an orthohexagonal unit cell to that of a primitive hexagonal one

$$c' = c, \ b' = b, \ a' = \frac{1}{2}(a + b).$$

The angle α' equals α, the angle γ' is found using elementary trigonometry, and β' comes from the formula connecting the cosines of the angles of vectors from the origin of an oblique coordinate system, giving: a' = 4.61; b' = 4.69; c' = 6.12 A; α' = 89°40; β' = 90°27, γ' = 59°12.

Delone's reduction [17] (with all obtuse angles) is achieved by the exchange of two of the three axes: a_m = -b', b_m = a', c_m = c', with a simultaneous exchange of the corresponding angles with their complements:

$$\alpha_m = \beta', \ \beta_m = 180° - \alpha', \ \gamma_m = 180° - \gamma'.$$

Nonetheless, all our later calculations were based on the nonprimitive unit cell C determined initially, since it is so close to pseudo-orthohexagonal, with $\alpha \approx \beta \approx \gamma \approx 90°$ and a : b $\approx \sqrt{3}$. More to the point is that under similar circumstances, the pseudohexagonal micas and other layer silicates would be described by a similar system, using the constants a and b for example.

The piezoelectric effect shows the absence of a center of symmetry (for the appropriate tests we thank the crystal physics faculty of Moscow State University). The center of symmetry problem has been solved more positively by a statistical analysis of the distribution of structure factors in a section of the reciprocal lattice [18].

The presence in the primitive unit cell of a single Rb atom, with a scattering ability greatly exceeding that of the other atoms, means that in comparisons with experimental results, we must use for the structure factor distribution function [18] not

$$_1N(z) = 1 - \exp z, \tag{1}$$

but the modification*

$$\max_1 N(z, r) = (1 + r^2) \exp(-r^2) \tag{2}$$

$$\times \int_0^z \left[\exp(-1 - r^2) z \right] I_0 [2r \sqrt{(1 + r^2) z}] \, dz.$$

Here $r = \dfrac{f_H}{\sqrt{\sum f_L^2}}$, and f_H and f_L are the scattering abilities of the heavy atom and light atoms, respectively, and I_0 is a first-order modified Bessel function.

With a single heavy atom in the unit cell, the origin may always be taken at this atom, and this was assummed in derivation of the last formula.

Since $f = f (\sin \vartheta / \lambda)$, then the parameter $\underline{r}$ may be found more exactly from

$$r = \frac{\int f_H d (\sin \vartheta / \lambda)}{\sqrt{\sum_j \int f_L^2 d(\sin^2 \vartheta / \lambda^2)}}$$

$$\approx \frac{\sum_k f_{Hk} \Delta (\sin \vartheta / \lambda)}{\sqrt{\sum_j \sum_k f_{jk} \Delta (\sin^2 \vartheta / \lambda^2)}}, \tag{3}$$

but in a visual estimation of intensities, $\underline{r}$ may be satisfactorily found from

$$r = \frac{z_H}{\sqrt{\sum z_L^2}} \tag{4}$$

(with a possible error of 10-15%).

In the integrand (2), we will expand the Bessel function $I_0(x)$ in a power series. Terminating the latter at any K > x/2, sufficiently large that the remaining terms

---

*As in [18], we will denote the distribution function for noncentrosymmetric structures by $_1N(z, r)$, and for centrosymmetric ones by $_{\bar{1}}N(z, r)$.

may be neglected, we have

$$\max{}_1 N(z, r) = \exp(-r^2)\Big\{[1 - \exp(-1 - r^2)z]$$
$$+ r^2[1 - (1 + (1 + r^2)z)\exp z(-1 - r^2)]$$
$$+ \frac{r^4}{4}[2 - ((1 + r^2)z^2$$
$$+ 2(1 + r^2)z + 2)\exp z(-1 - r^2)] \qquad (5)$$
$$+ \frac{r^6}{36}[6 + (-z^3(1 + r^2)^3 - 3z^2(1 + r^2)^2$$
$$- 6z(1 + r^2) - 6)\exp z(-1 - r^2)] + \cdots\Big\}.$$

Having the equation of the surface $f(r, z)$, we can derive any of its sections, the curve $N(z)$, for any specific value of r in the structure. The results of our calculations, using the comparatively simple formula (5), correspond with three-figure accuracy with the results in [19], which were derived by numerical integration. We thought this mathematical digression was of value in view of the fact that structure analysts are going over more and more to statistical analysis of intensities.

On analysis of the material on Rb difluoroberyllate, the experimental points $N(r, z)$ fall for the two zones $h0l$ and $0kl$ (Fig. 1) in a sufficiently narrow region be-

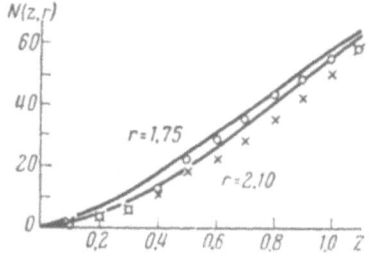

Fig. 1. Comparison of the experimental distribution for $RbBe_2F_5$ with the theoretical distributions $_1N(z, 1.77)$ and max $_1N(z, 2.1)$. $\times$ = $h0l$ zone, $\bigcirc$ = $0kl$ zone.

tween the $_1N(r, z)$ curves (both for the noncentrosymmetric case), of which one corresponds to r = 2.10, found using (4), and the other to r = 1.77 from (3). Thus, the statistical analysis of intensities leads fairly conclusively to the single group P1(C1 in the arrangement we have taken).

The noncentrosymmetric distribution of the atoms leads to errors on the Patterson synthesis diagram; the spread-out peaks in both P(xz) and P(yz) projections do not accurately represent the heights of the maxima, and do not allow immediate visual determination of the coordinates of specific peaks.

The successful use of the method of integral characteristics [20, 21] in our previous analysis of the structure of lovozerite, also noncentrosymmetric, prompted us to apply the method in this case as well.

For a positive determination of the points of emergence of specific Patterson peaks on the two projections, the absolute heights of the maxima and their central points were calculated theoretically, using

$$p_{ij}(0) = \frac{1}{2\pi}\int_0^\infty f_i(s)f_j(s)\cdot s\,ds, \qquad (6)$$

and also the shapes of the maxima (with specially constructed strips)

$$p_{ij}(r) = \frac{1}{2\pi}\int_0^\infty f_i(s)f_j(s)sI_0(sr)\,ds \qquad (7)$$
$$\approx \frac{1}{2\pi}\sum_{k=1}^{k_{max}} f_i(s_k)f_j(s_k)s_kI_0(s_kr)\Delta s_k.$$

This was done for the peaks Rb-Rb, Rb-F, F-F, Rb-Be, F-Be, and Be-Be, at a value of the temperature factor of $B_{0kl} = B_{h0l} = 2A^2$.

The experimental heights of the Patterson peaks were converted to the absolute scale using

$$p_{abs} = Kp_{exp} \div \frac{F_{000}^2}{S}, \qquad (8)$$

in which K is determined from the relation

$$Kp(00)_{exp} = \sum_{i=1}^N p_{ii}(0) \div \frac{F_{000}^2}{S}.$$

$p(00)_{exp}$ is the value of the nonnormalized function at the origin of the Patterson synthesis, $F_{000}^2 = \sum_{i=1}^N Z_i^2$, S is the area of the projection, $p_{ii}(0)$ is calculated from (6).

The close correspondence in height and shape of the experimental and calculated Patterson peaks meant that we could determine the approximate coordinates of Rb and three F atoms and construct a rough model of the structure. The coordinates of atoms F_3 and F_4 were determined from a modified Patterson synthesis starting at $(F_{exp} - F_{Rb})^2$.

The next Patterson synthesis was again a difference synthesis of $F_{exp}^2 - F_H^2$, where F_H was determined by Rb and F. In it, it was necessary to distinguish between three types of peak, 2RbBe, 2FBe, and Be^2, where the peaks corresponding to the pair Rb-Be markedly predominated over the rest. Use of the method of integral characteristics [19] allowed us here also to separate and localize sufficiently accurately Be_I and Be_{II}.

In electron density syntheses the early stages of refining the coordinates were carried through without taking into account the Be atom ($2Z_{Be}/Z_{Rb} + 5Z_F \approx 8/82 \approx 10\%$), and their contributions were only introduced into the structure factors when the coefficient of divergence R had reached a value of 13-14%.

The difference synthesis starting with $F_{exp} - F_H$ (Rb + 5F) confirmed the coordinates of F, but somewhat displaced the Be atoms and allowed a correction to be made in the temperature factor for atoms F_1 and F_5[†].

In the well-known formula [22] which gives ΔB_j for the jth atom,

$$\Delta B_j = \frac{\left(\frac{\partial^2 R}{\partial x^2}\right)_j}{\frac{\partial}{\partial B}\left(\frac{\partial^2 R}{\partial x^2}\right)} = \frac{\left(\frac{\partial^2 R}{\partial x^2}\right)_j}{\frac{\partial}{\partial B}\left(\frac{\partial^2 \rho_\theta}{\partial x^2}\right)} = \frac{\left(\frac{\partial^2 R}{\partial x^2}\right)_j}{\frac{\partial}{\partial B}\left(\frac{\partial^2 \sigma}{\partial x^2}\right)},$$

(10)

the numerator comes directly out of the synthesis, and to find the denominator the method of integral characteristics was again used successfully:

$$\left(\frac{\partial^2 \sigma}{\partial x^2}\right)_0 = \frac{\partial}{\partial x^2}\left\{\frac{1}{2\pi}\int f(s)\, sI_0(sr)\, ds\right\}$$

$$= -\frac{1}{4\pi}\int j(s)\, s^3\, ds$$

(11)

and

$$\frac{\partial}{\partial B}\left(\frac{\partial^2 \sigma}{\partial x^2}\right) = -\frac{1}{4\pi}\int s^3\, ds\, \frac{df}{dB}$$

$$= \frac{1}{64\pi^3}\int fs^5\, ds \approx \sum_{k=0}^{K} D(s_k)\, ds,$$

(12)

where $D(s) = \frac{fs^5}{64\pi^3}$ is the integrand. From this we find that $\Delta B = 0.1\ \text{A}^2$, and

$$B_{F_1} = B_{F_5} = 2.1\ \text{A}^2.$$

Introduction of the true (isotropic) temperature factor for atoms F_1 and F_5 and refining of the coordinates of all atoms leads to a decrease in the coefficient of divergence to 11.5 and 11.8% respectively for R_{h0l} (3 zero reflections) and R_{0kl} (3 zero reflections) up to $\sin \vartheta/\lambda = 0.8\ \text{A}^{-1}$[‡].

The coordinates of the basal atoms which were derived from the synthesis after reaching these coefficients of divergence are given in Table 1. They include twen-

TABLE 1. Coordinates of Basal Atoms in the Structure of Rubidium Di(Meta) Fluoroberyllate

Atom	x	y	z
Rb	0	0	0
F_1	16.7	49.4	17.5
F_5	31.9	0.4	80.0
F_2	—2.5	49.5	47.0
F_3	20.0	83.4	46.7
F_4	25.0	33.4	52.5
Be_1	15.0	54.5	41.3
Be_{11}	31.0	5.0	56.3

*In hundredths parts of unit cell sides.

[†] Having the least symmetrical environment, cf. Table 2.
[‡] The maximum values of $\sin \vartheta/\lambda$ were greatly compressed (large temperature factor) along the axis $\vartheta = 0$ on our Mo Weissenberg photos.

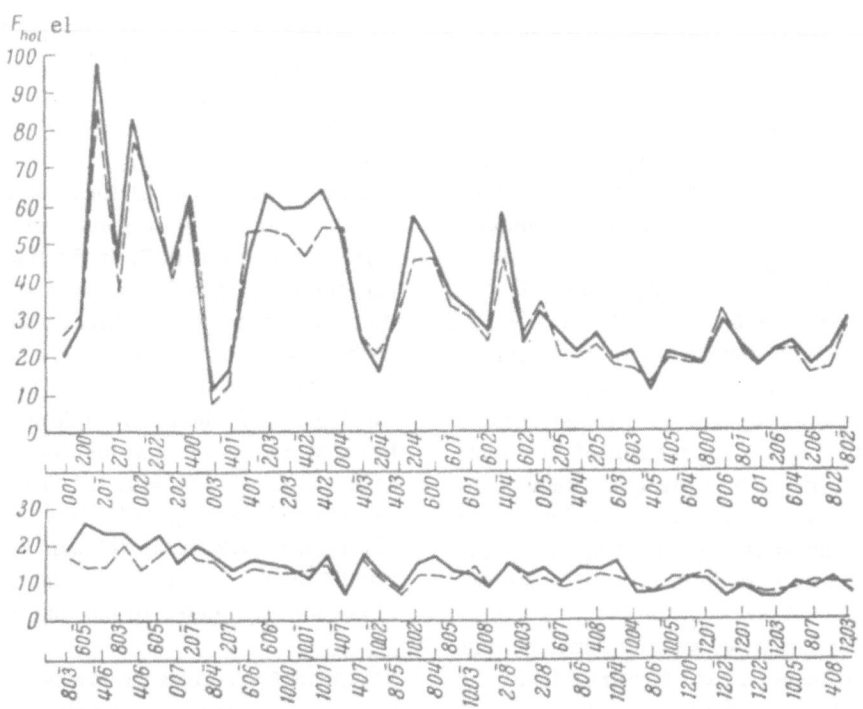

Fig. 2. Structure amplitudes for the h0l zone. Comparison of calculation (solid line) and experiment (dashed line).

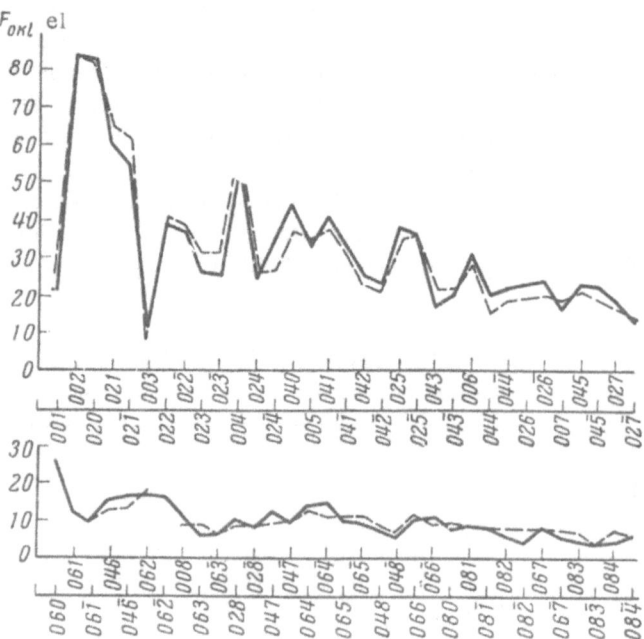

F_{okl}, el

Fig. 3. Structure amplitudes for the 0kl zone. Comparison of calculation (solid line) and experiment (dashed line).

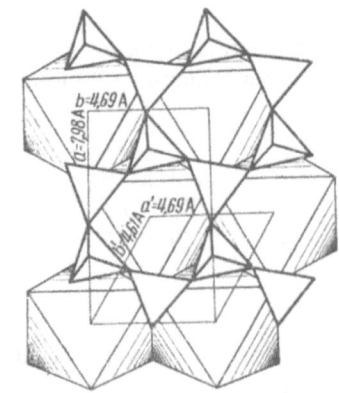

Fig. 4. Structure of RbBe$_2$F$_5$ as polyhedra. xy projection. Rb atoms in the octahedra, Be atoms in the tetrahedra.

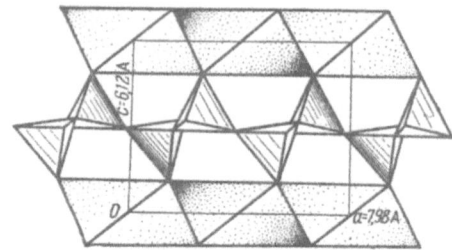

Fig. 5. Structure of RbBe$_2$F$_5$. xz projection. Alternation of brucite-type layers of squashed Rb octahedra and double networks of Be tetrahedra.

ty-one parameters. The accuracy of the determination [23] is somewhat different on the two projections:

$h0l$: Rb $\pm$ 0.002   F $\pm$ 0.01   Be $\pm$ 0.03A
$0kl$: Rb $\pm$ 0.0025   F $\pm$ 0.015   Be $\pm$ 0.04A

The experimental structure amplitudes, converted to the absolute scale, and the amplitudes calculated from the coordinates in Table 1 are compared graphically in Figs. 2 and 3.

In the xz projection (Fig. 4) of the structure of Rb dimetafluoroberyllate, similar to those of talc and phlogopite, we see a continuous brucite-type layer [24] constructed of compressed RbF$_6$ octahedra, with almost ideal hexagonal symmetry. The distance between identical points above and below (Fig. 5) is the lattice constant c = 6.12 A. Between the layers of Rb octahedra is a double network of BeF$_4$ tetrahedra. While, however, the Rb octahedron layer is continuous, we see on examining the projection (Fig. 5) that the Be tetrahedron layer is not continuous, but is made up of six-membered (more strictly, ditrigonal) rings. Along the axis of the pseudo-hexagonal unit cell, in bands of triangles, two occupied positions alternate with one vacant, so that the ratio of the number of F atoms at the vertices of the triangular bases of the Be tetrahedra to the number of Be atoms is 3 : 2.

However, as well as these, we must remember that above or below the center of each Be tetrahedron there is another atom of F, and so the formula of the fluoroberyllate tetrahedron layer will not be Be$_2$F$_3$, as in the ring network of octahedra in the corundum-muscovite

Al$_2$O$_3$ network [24], but Be$_2$F$_5$. The network is taken as infinite in two directions, so we will express it by the formula [Be$_2$F$_5$]$_{\infty, \infty}$ This corresponds exactly with the silicon-oxygen networks in the best-known dimetasilicates, i.e., talc, micas, etc., which have an infinite two-dimensional network of [Si$_2$O$_5$]$'_{\infty, \infty}$** units, but the resemblance breaks down in the other projection. In these silicates all the tetrahedra have their vertices pointing in one direction, along the z axis toward one layer of octahedra, while in the fluoroberyllate half the BeF$_4$ tetrahedra point to one side along the z axis, toward one layer of Rb octahedra, and the other half point to the other side, toward the next layer of octahedra. Each network, therefore, belongs not to a single layer of octahedra, as in the comparable silicates, but to two layers at once, serving as the link between them. We have said that the network is "double," but if we compare (Fig. 5) the height of a layer of tetrahedra with that of a layer of octahedra, we could say it was a "one-and-a-half" layer. In symmetry terms we could say that if in the dimetasilicates all the rings are characterized by

---

**Si$_2$O$_5$ or Be$_2$F$_5$ gives the average number of tetrahedra or anions making up, on average, one ring, if we take into account that each tetrahedron is shared between three six-membered rings.

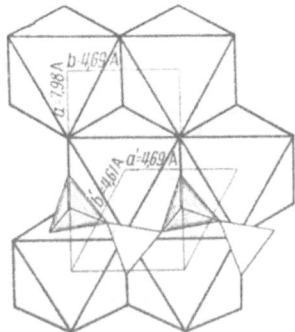

Fig. 6. Pyroxene-type $[Be_{1+1}F_6]_\infty$ chains, selected from the fluoro-beryllate network in $RbBe_2F_5$. In the two-membered link, neighboring Be tetrahedra have their free vertices pointing in opposite directions.

pseudosymmetry with a sixfold symmetry axis, then in Rb fluoroberyllate there are dioptase-type rings [25] with a sixfold axis of rotation, $6 ° = \bar{3}$; three tetrahedra in each ring point to one side, and three to the other, each set of three pointing to a different octahedron layer. In the fluoroberyllate there is no layer of vacant octahedra, like that which separates the three-layer packets in talc, nor a layer of very sparsely occupied octahedra with K, Na, or Ca, as in the micas.

The perfect cleavage of the difluoroberyllate is, therefore, not due to layers of vacant octahedra, as in the micas and chlorites, but is due to the lack of "co-hesion" of the Rb cation in the octahedron, due to its small charge and large radius. The inference that cleav-age is really taking place through the octahedra follow from the striations on the (001) cleavage plane, corre-sponding to "columns" of octahedra, on which the bru-cite-type layers cleave easily (Fig. 4).

Up to now we have compared the structure of the fluoroberyllate with the structures of the better-known layer silicates and noted the great similarity in their "architecture": layers of octahedra, alternating with layers of tetrahedron rings. The fact that the latter are "one-sided" in silicates and "two-sided" (reversing) in the fluoroberyllate does not seem to make a great deal of difference. The latter fact, incidently, is clearly seen if we compare the formulas and observe that in the Mg silicates there must be three Mg octahedra to one oxy-gen-silicon ring, while in our fluoroberyllate there must be only one Rb octahedron to one ring of BeF_4 tetrahedra. This is not unexpected if we remember the large size of the Rb octahedron (ionic radius of Rb is 1.49 A, of Mg 0.78 A). Networks of BeF_4 tetrahedra in one case and SiO_4 tetrahedra in the other, both of about the same size, cannot match up in identical fashion with close-packed octahedron layers of differing dimensions. In the Mg silicates, every tetrahedron of the network is connected (all on one side) to the vertices of Mg octahedra, while in the fluoroberyllate only half the Be tetrahedra are con-nected (on any one side) to the vertices of Rb octahedra.

A network of oxygen-silicon tetrahedra is often re-presented as a result of condensation of pyroxene chains $[SiO_3]_\infty$, through amphibole bands $[Si_4O_{11}]_\infty$, to two-di-mensional networks $[Si_8O_{20}]_{\infty,\infty}$ If we select pyroxene-type chains of $[BeF_3]_\infty$ from the fluoroberyllate network (Fig. 6), we see that the tetrahedra making up these chains alternately face different sides of the plane of the network, unlike the silicate chains, where all the tetrahedra face one side. We can express this in another way by saying that if in the Mg silicates the side of an Mg octahedron matches up (more or less exactly) with the side of an Si tetrahedron, then in our fluoroberyllate the side of an RB octahedron, equal to $b = 4.69$ A, matches up with two Be tetrahedra (or one diortho group Be_2O_7) selected from the $[Be_2F_5]_{\infty,\infty}$ network. Rb diflu-oroberyllate is therefore a model for silicates from the "Second Section of silicate crystal chemistry," those to which the majority of this laboratory's papers have been devoted in recent years [26], in which the elementary building brick is the diortho group Si_2O_7, and not the SiO_4 tetrahedron, as in the "First Section."

For the pseudohexagonal structure of $RbBe_2F_5$ (actu-ally triclinic), with two sorts of Be atoms and five sorts of F atoms, we can derive the ideal valence balance (Table 2), fully comparable with that of beryl with its true hexagonal structure. In Table 3, the interatomic distances are given separately for each sort of atom. In the undisputed BeF_4 tetrahedron, both the Be–F and the F–F distances fall within narrow limits, 1.43-1.48 and 2.33-2.41 A. In the octahedron round the "poorly-co-hesive" Rb atom, the Rb–F distance also is fixed be-tween the narrow limits 2.82-3.02 A (sum of radii, 2.82 A) but the F–F distances clearly fall into two groups. Like the Ca octahedra in milarite [27], the Rb octahedra are very squashed, so that at the bottom of the octahedra we have F–F = 4.61-4.69 A, and at the side edges 3.55-3.60 A, so within each group the spread of results is very small. The physical properties of the artificial crystals, and their relation to structure, are less interesting than the well-documented properties of the natural minerals. We have dealt with cleavage and its structural basis above.

TABLE 2. Valence Balance in the $RbBe_2F_5$ Structure

Cations \ Anions	Rb	Be_I	Be_{II}	$\sum \frac{\omega_i}{n_i}$
F_1	$3 \times 1/6$	1/2	—	1
F_2	—	1/2	1/2	1
F_3	—	1/2	1/2	1
F_4	—	1/2	1/2	1
F_5	$3 \times 1/6$	—	1/2	1

TABLE 3. Interatomic Distances in $RbBe_2F_5$ (in A)

Rb Octahedra

$Rb - F_1 - 2.87$	$F_1 = F_1{}^* = 4.61$	$F_1 - F_5 = 3.55$
$F_1{}^* = 2.92$	$F_1 - F_1{}^{**} = 4.69$	$- F_5{}^* = 3.58$
$F_1{}^{**} = 2.85$	$F_1{}^* - F_1{}^{**} = 4.61$	$F_1{}^* - F_5{}^* = 3.60$
$F_5 = 2.82$	$F_5 - F_5{}^{**} = 4.61$	$F_1{}^* - F_5{}^{**} = 3.55$
$F_5{}^* = 3.02$	$F_5 - F_5{}^* = 4.61$	$F_1{}^{**} - F_5 = 3.60$
$F_5{}^{**} = 2.99$	$F_5{}^* - F_5{}^{**} = 4.69$	$- F_5{}^{**} = 3.58$

Be_I Tetrahedra (A)

$Be_1 - F_1 = 1.48$	$F_1 - F_2 = 2.35$	$F_2 - F_3 = 2.39$
$F_2 = 1.46$	$F_1 - F_3 = 2.41$	$- F_5 = 2.34$
$F_3 = 1.45$	$- F_4 = 2.37$	$F_3 - F_4 = 2.40$
$F_4 = 1.44$		

Be_{II} Tetrahedra (A)

$Be_{11} - F_5 = 1.47$	$F_5 - F_2{}^* = 2.38$	$F_2{}^* - F_3 = 2.33$
$F_2{}^* = 1.45$	$- F_3 = 2.38$	$- F_4 = 2.41$
$F_3 = 1.46$	$- F_4 = 2.36$	$F_2 - F_4 = 2.40$
$F_4 = 1.43$		

*/** denotes, as usual, nonbasal atoms, either related to the basal atoms by a translation of c/2, or from other unit cells.

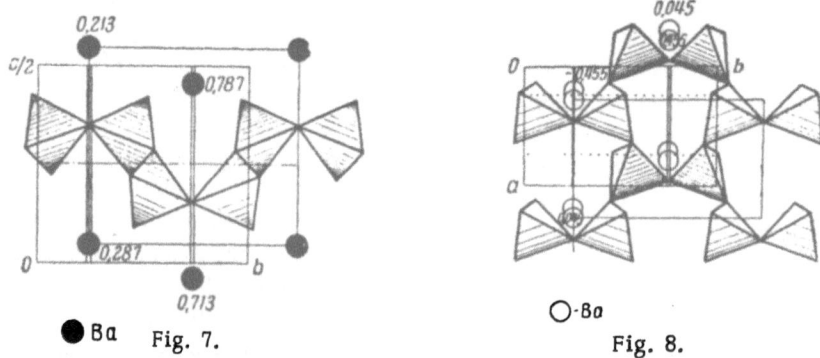

Fig. 7.   Fig. 8.

Fig. 7. Pyroxene-type chains $[Si_{2+2}O_{12}]$ selected from the oxygen-silicon network in sanbornite, $BaSi_2O_5$. All the links of the chains are diortho groups, Si_2O_7.

Fig. 8. The six-membered rings from which the oxygen-silicon network is constructed in sanbornite. In each ring four of the tetrahedra are orientated in the reverse direction to the other two.

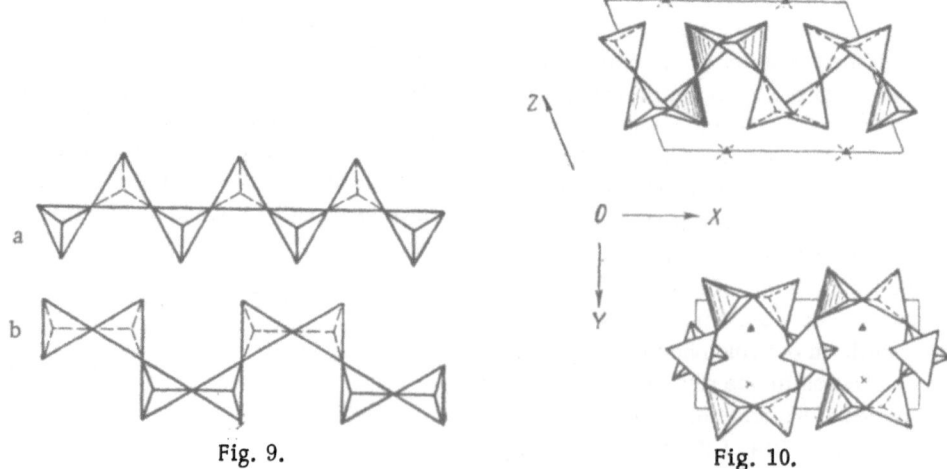

Fig. 9.   Fig. 10.

Fig. 9. The two types of oxygen-silicon chain in sanbornite with the single general formula $[SiO_3]_\infty$. a) Pyroxene-type $[Si_{1+1}O_6]_\infty$ chains; b) narsarsukite-type $[Si_{2+2}O_{12}]_\infty$ chains.

Fig. 10. Petalite. Amphibole-type bands with deep corrugations, selected from the oxygen-silicon network. Projections on the two planes xy and xz.

The double (one-and-a-half) thickness network is able to connect with equal success layers of octahedra both by retaining the angles α and β and by replacing these angles by their complements, i.e., creating twin bands which give dense polysynthetic intergrowths. Analogy with layer silicates leaves no doubt that the fluoroberyllate crystals are optically negative.

In the preceding account we have compared the first dimetafluoroberyllate which has been studied in detail with the only classical layer dimetasilicates well-known to Goldschmidt and Fersman. A number of similar features have been shown, and also some differences. The basic difference is that these dimetasilicates, mainly Mg (Fe) and Al silicates, are from the "First Section" of silicate crystal chemistry, while Rb di(meta)fluoroberyllate can only serve as a model for silicates from the "Second Section." In 1958 an interpretation appeared for the layer silicate sanbornite [13], which undoubtedly belongs to the second section, and, in fact, in agreement with Goldschmidt's predictions, the formula of this silicate, $BaSi_2O_5$, is completely analogous to that of the Rb difluoroberyllate: the $[Be_2F_5]^1$ radical corresponds to $[Si_2O_5]^2$, the large Rb^{1+} cation corresponds to the large Ba^{2+} ion, exactly according to Goldschmidt. The arrangement is not, however, repeated exactly. Although the radius of Ba^{2+} is less than that of Rb^{1+}, the Ba cation is situated in a larger polyhedron, namely a heptaverticon, similar to that frequently found in the coordination of Ca (sphene, epidote). The heptavertica are also packed in a compressed layer, parallel to (001), but this layer is corrugated. Polysynthetic twinning in Rb fluoroberyllate corresponds with "twinning" within the unit cell itself in the Ba silicate (Ito, [28]), where within the c lattice distance there are two corrugated layers with a

separation of $c/2 = 6.77$ A, a little more than that in the fluoroberyllate.

As in Rb fluoroberyllate, in sanbornite there is a double (one-and-a-half) thickness network of $[Si_2O_5]_{\infty,\infty}$ radicals between consecutive layers of Ba polyhedra.

In this Ba silicate, undoubtedly belonging to the second section, the basic building blocks of the oxygen-silicon arrangement are clearly-resolved $[Si_2O_7]$ units; along the x axis they are completely discrete (Fig. 7) and interconnected by the other $[Si_2O_7]$ units in the other portion of the multiple layer. Along the y axis, however, we again see chains constructed on the lines of Rb fluoroberyllate; the repeat unit Si_2O_7 is made up of one tetrahedron pointing upward along the z axis, and another pointing downwards (Fig. 8). We can, therefore, select metasilicate chains with the single general formula $[SiO_3]_\infty$ from the oxygen-silicon network in sanbornite, but along the x axis direction this will be $[Si_{1+1}O_6]_\infty$, i.e., pyroxene-type chains with Si tetrahedra pointing in different directions (Fig. 9a), while along the y axis these $[Si_{2+2}O_{12}]_\infty$ chains are of the narsarsukite type (Fig. 9b). The $[Si_2O_5]_{\infty,\infty}$ network itself is made up of six-membered rings, like those in talc, micas, etc., and also in the fluoroberyllate, but in each ring four tetrahedra point in one direction and two in the other. In the pseudo-orthohexagonal network of these rings, the two rings in the unit cell are related by an oblique glide plane, i.e., if in one of them four tetrahedra are pointing upwards, then in the other, four will be pointing downwards (Fig. 8).

With the structures of $RbBe_2F_5$ and $BaSi_2O_5$ we may compare some structures calculated in recent years for petalite ($LiAlSi_4O_{10}$) [29, 30], α-$Na_2Si_2O_5$ and $Li_2Si_2O_5$. All these layer structures are like the first two in that the layers are not of simple polyhedra, but have double-

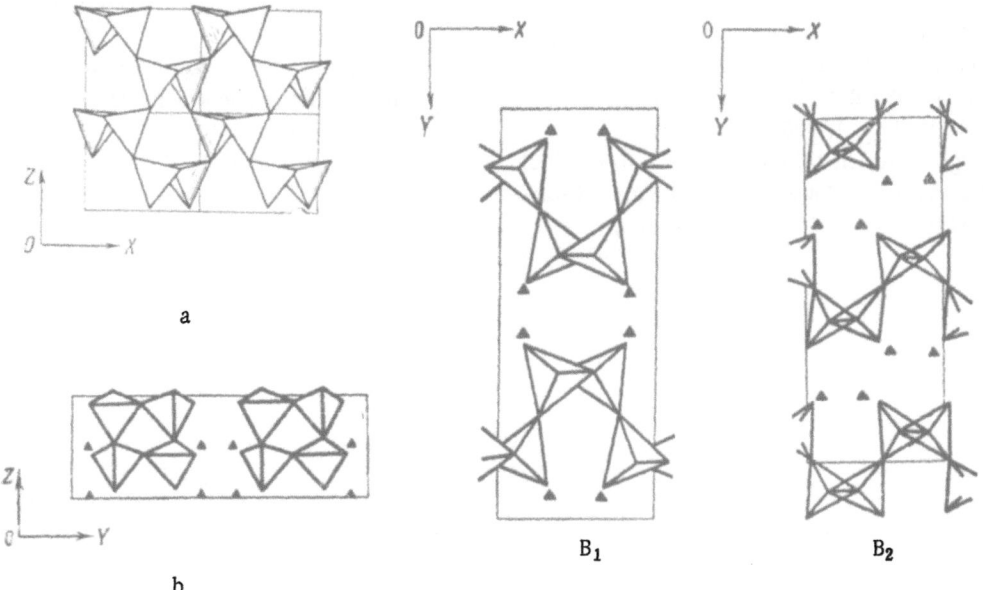

Fig. 11. The oxygen-silicon network in $Li_2Si_2O_5$ and in α-$Na_2Si_2O_5$. a and b are similar projections on xz and zy; B_1 and B_2 are slightly differing projections on yz.

sided $[Si_2O_5]_{\infty, \infty}$ oxygen-silicon networks composed of six-membered rings, connecting two layers of cation polyhedra. However, one of the main features of these oxygen-silicon networks is their intense corrugation, making it difficult to allocate them definitely to either first or second section, especially since the Al and Li cations are situated in tetrahedra instead of octahedra.

In the oxygen-silicon arrangement in petalite, the basis is a pyroxene-type chain as in $RbBe_2F_5$, i.e., reversing (Fig. 10). Two such chains are connected into a band, as in the amphiboles, not by a symmetry plane like the amphiboles, but by a diad, giving rings similar to those in dioptase. Neighboring bands interlock obliquely in such a way that every other tetrahedron from the upper half of the band interlocks with a second tetrahedron from the lower half, and so on (Fig. 11a). The extreme corrugation is seen in Fig. 11b. In $\alpha\text{-}Na_2Si_2O_5$ and $Li_2Si_2O_5$ the corrugation reaches such a degree that in each hexagonal ring in the $[Si_2O_5]_{\infty, \infty}$ network, one half of the ring points up and the other half down, i.e., the $[SiO_3]_{\infty}$ chain is really of the pyroxene type, and the corresponding silicates should be allocated to the first section of silicate crystal chemistry.

LITERATURE CITED

1. T. Hahn, Neues J. Mineral. 86, 1, 1-65 (1953).
2. N. A. Toropov and I. A. Bondar', Usp. Khimii 24, 52 (1955).
3. A. V. Novoselova, Usp. Khimii 28, 33 (1959).
4. R. G. Grebenshchikov, Doklady Akad. Nauk SSSR 114, 316 (1955).
5. N. A. Toropov and R. G. Grebenshchikov, Zhur. Neorgan. Khimii 1, 7, 1619 (1956).
6. A. V. Novoselova and Yu. P. Simanov, Uch. Zap. MGU 174, 7 (1955).
7. N. A. Toropov and R. G. Grebenshchikov, Zhur. Neorgan. Khimii 1, 12, 2686 (1956).
8. V. M. Goldschmidt, Skrifter Norske Videnskapa. Akad., Oslo, Mat.-Nat. KL 7, 1, (1926); 8, 8a, 7 (1926).
9. V. M. Goldschmidt, Geochem. Verteilungsgesetze 8, 127-139 (Oslo, 1927).
10. A. V. Novoslova, M. E. Levina, Yu. P. Simanov, and A. G. Zhasmin, Zhur. Organ. Khimii 14, 6, 385 (1944).
11. A. V. Novoselova, Yu. P. Simanov, and E. I. Yarmbash, Zhur. Fiz. Khimii 26, 9, 1245 (1952).
12. N. A. Toropov and R. G. Grebenshchikov, Zhur. Neorgan. Khimii 6, 4, 920 (1961).
13. R. M. Douglass, Amer. Mineralogist 43, 517 (1958).
14. O. N. Breusov, A. V. Novoselova, and Yu. P. Simanov Doklady Akad. Nauk SSSR 118, 935 (1958).
15. L. F. Kirkina, A. V. Novoselova, and Yu. P. Simanov Zhur. Neorgan. Khimii 1, 125 (1956).
16. L. M. Mikheeva, A. V. Novoselova, and R. S. Baktimirov, Zhur. Neorgan. Khimii 1, 499 (1956).
17. B. N. Delone, Zeit. Kristallographie 84, 132 (1933)
18. E. R. Howells, Acta Crystallographica 3, 210 (1950).
19. G. A. Sim, Acta Crystallographica 4, 123 (1958).
20. V. V. Ilyukhin and S. V. Borisov, Zhur. Strukt. Khimii 1, 1, 80 (1960).
21. V. V. Ilyukhin and N. V. Belov, Kristallografiya 5, 2, 200 (1960) [Soviet Physics — Crystallography, Vol. 5, p. 186].
22. H. Lipson and V. Cochrane, Determination of the Structure of Crystals [Russian translation] (Moscow, IL, 1956).
23. B. K. Vainshtein, Zhur. Éksperim. i Teor. Fiz. 27, 44 (1954).
24. N. V. Belov, V. P. Butuzov, and N. I. Golovastniko Doklady Akad. Nauk SSSR 87, 6 (1953).
25. N. V. Belov, Structure of Ionic Crystals [in Russian] (Moscow, 1947).
26. N. V. Belov, The Crystal Chemistry of Silicates w' Large Cations (The second of the V. I. Vernadskii Lectures) (Moscow, Acad. Sci. USSR Press, 1961).
27. N. V. Belov and T. N. Tarkhova, Doklady Akad. Nauk SSSR 69, 3, (1949). Trudy Institute Kristallografiya, Akad. Nauk SSSR 6 83-140 (1951).